A REVIEW OF THE DEPARTMENT OF ENERGY CLASSIFICATION POLICY AND PRACTICE

Committee on Declassification of Information for the Department of Energy Environmental Remediation and Related Programs

Board on Radioactive Waste Management

Commission on Geosciences, Environment, and Resources

National Research Council

NATIONAL ACADEMY PRESS
WASHINGTON, DC 1995

NOTICE: The project that is the subject of this report was approved by the Governing board of the National Research Council, whose members are drawn from the councils of the National Academy of Sciences, the National Academy of Engineering, and the Institute of Medicine. The members of the committee responsible for the report were chosen for their special competences and with regard for appropriate balance.

This report has been reviewed by a group other than the authors according to procedures approved by a Report Review Committee consisting of members of the National Academy of Sciences, the National Academy of Engineering, and the Institute of Medicine.

Support for this study on Declassification of Information for the Department of Energy Environmental Remediation and Related Programs was provided by the U.S. Department of Energy, under agreement DE-FC01-94EW54069.

Library of Congress Catalog Card No. 95-70717
International Standard Book Number 0-309-05338-2

Additional copies of this report are available from:

National Academy Press
2101 Constitution Avenue, N.W.
Box 285
Washington, D.C. 20055
800-624-6242
202-334-3313 (in the Washington Metropolitan Area)

B-666

Copyright 1995 by the National Academy of Sciences. All rights reserved.

Cover design by Terri M. Jackson
Printed in the United States of America

COMMITTEE ON DECLASSIFICATION OF INFORMATION FOR THE DEPARTMENT OF ENERGY ENVIRONMENTAL REMEDIATION AND RELATED PROGRAMS

RICHARD A. MESERVE, *Chair*, Covington & Burling, Washington, D.C.
DEAN E. ABRAHAMSON, University of Minnesota, Minneapolis
LYNDA L. BROTHERS, Davis Wright Tremaine, Seattle, Washington
THOMAS A. COTTON, JK Research Associates, Inc., Arlington, Virginia
PAUL P. CRAIG, University of California, Davis
GEORGE A. FERGUSON, Howard University (ret.), Washington, D.C.
H. JACK GEIGER, City University of NY Medical School, New York, New York
MICHELE S. GERBER, Westinghouse Hanford Company, Richland, Washington
KONRAD B. KRAUSKOPF, Stanford University, Stanford, California
WOLFGANG K. H. PANOFSKY, Stanford University, Stanford, California
RICHARD B. SETLOW, Brookhaven National Laboratory, Upton, New York
PATRICIA A. WOOLF, Princeton University, Princeton, New Jersey

Staff

CARL A. ANDERSON, Staff Director
ROBERT S. ANDREWS, Senior Staff Officer
DENNIS L. DuPREE, Senior Project Assistant
JO L. HUSBANDS, Senior Staff Officer

BOARD ON RADIOACTIVE WASTE MANAGEMENT

MICHAEL C. KAVANAUGH, *Chair*, ENVIRON Corporation, Emeryville, California
B. JOHN GARRICK, *Vice-Chair*, PLG, Incorporated, Newport Beach, California
JOHN F. AHEARNE, Sigma Xi, The Scientific Research Society, Research Triangle Park, North Carolina
JEAN M. BAHR, University of Wisconsin, Madison
LYNDA L. BROTHERS, Davis Wright Tremaine, Seattle, Washington
SOL BURSTEIN, Wisconsin Electric Power, Milwaukee (retired)
MELVIN W. CARTER, Georgia Institute of Technology, Atlanta (emeritus)
PAUL P. CRAIG, University of California, Davis (emeritus)
MARY R. ENGLISH, University of Tennessee, Knoxville
ROBERT D. HATCHER, JR., University of Tennessee/Oak Ridge National Laboratory, Knoxville
DARLEANE C. HOFFMAN, Lawrence Berkeley Laboratory, Berkeley, California
H. ROBERT MEYER, Keystone Scientific, Inc., Fort Collins, Colorado
PERRY L. McCARTY, Stanford University, Stanford, California
CHARLES McCOMBIE, National Cooperative for the Disposal of Radioactive Waste, Wettingen, Switzerland
PRISCILLA P. NELSON, Universtiy of Texas at Austin
D. KIRK NORDSTROM, U.S. Geological Survey, Boulder, Colorado
D. WARNER NORTH, Decision Focus, Inc., Mountain View, California
GLENN PAULSON, Illinois Institute of Technology, Chicago
PAUL SLOVIC, Decision Research, Eugene, Oregon
BENJAMIN L. SMITH, Independent Consultant, Columbia, Tennessee

Staff

CARL A. ANDERSON, Staff Director
KEVIN D. CROWLEY, Associate Director
ROBERT S. ANDREWS, Senior Staff Officer
KARYANIL T. THOMAS, Senior Staff Officer
THOMAS KIESS, Staff Officer
SUSAN B. MOCKLER, Research Associate

ROBIN L. ALLEN, Senior Project Assistant
REBECCA BURKA, Senior Project Assistant
LISA J. CLENDENING, Senior Project Assistant
DENNIS L. DuPREE, Senior Project Assistant
SCOTT HASSELL, Project Assistant
PATRICIA A. JONES, Project Assistant

COMMISSION ON GEOSCIENCES, ENVIRONMENT, AND RESOURCES

M. GORDON WOLMAN, *Chair,* The Johns Hopkins University, Baltimore, Maryland
PATRICK R. ATKINS, Aluminum Company of America, Pittsburgh, Pennsylvania
EDITH BROWN WEISS, Georgetown University Law Center, Washington, D.C.
JAMES P. BRUCE, Canadian Climate Program Board, Ottawa, Ontario, Canada
WILLIAM L. FISHER, University of Texas at Austin
EDWARD A. FRIEMAN, University of California, La Jolla
GEORGE M. HORNBERGER, University of Virginia, Charlottesville
W. BARCLAY KAMB, California Institute of Technology, Pasadena
PERRY L. MCCARTY, Stanford University, Stanford, California
S. GEORGE PHILANDER, Princeton University, Princeton, New Jersey
RAYMOND A. PRICE, Queen's University at Kingston, Ontario, Canada
THOMAS C. SCHELLING, University of Maryland, College Park
ELLEN SILBERGELD, Environmental Defense Fund, Washington, D.C.
STEVEN M. STANLEY, The Johns Hopkins University, Baltimore, Maryland
VICTORIA J. TSCHINKEL, Landers and Parsons, Tallahassee, Florida

Staff

STEPHEN RATTIEN, Executive Director
STEPHEN PARKER, Associate Executive Director
MORGAN GOPNIK, Assistant Executive Director
JIM MALLORY, Administrative Officer
SANDI FITZPATRICK, Administrative Associate

The National Academy of Sciences is a private, nonprofit, self-perpetuating society of distinguished scholars engaged in scientific and engineering research, dedicated to the furtherance of science and technology and to their use for the general welfare. Upon the authority of the charter granted to it by the Congress in 1863, the Academy has a mandate that requires it to advise the federal government on scientific and technical matters. Dr. Bruce Alberts is president of the National Academy of Sciences.

The National Academy of Engineering was established in 1964, under the charter of the National Academy of Sciences, as a parallel organization of outstanding engineers. It is autonomous in its administration and in the selection of its members, sharing with the National Academy of Sciences the responsibility for advising the federal government. The National Academy of Engineering also sponsors engineering programs aimed at meeting national needs, encourages education and research, and recognizes the superior achievements of engineers. Dr. Harold Liebowitz is president of the National Academy of Engineering.

The Institute of Medicine was established in 1970 by the National Academy of Sciences to secure the services of eminent members of appropriate professions in the examination of policy matters pertaining to the health of the public. The Institute acts under the responsibility given to the National Academy of Sciences by its congressional charter to be an adviser to the federal government and, upon its own initiative, to identify issues of medical care, research, and education. Dr. Kenneth Shine is president of the Institute of Medicine.

The National Research Council was organized by the National Academy of Sciences in 1916 to associate the broad community of science and technology with the Academy's purposes of furthering knowledge and of advising the federal government. Functioning in accordance with general policies determined by the Academy, the Council has become the principal operating agency of both the National Academy of Sciences and the National Academy of Engineering in providing services to the government, the public, and the scientific and engineering communities. The Council is administered jointly by both Academies and the Institute of Medicine. Dr. Bruce Alberts and Dr. Harold Liebowitz are chairman and vice chairman, respectively, of the National Research Council.

PREFACE

On December 7, 1993, Secretary of Energy Hazel O'Leary announced that the Department of Energy (DOE) had commenced aggressive efforts "to lift the veil of Cold War secrecy" that has surrounded many of DOE's activities. She declared her intention to declassify significant amounts of information that had previously been withheld from the public for reasons of national security. This report has its origin in the Secretary's subsequent request to the National Academy of Sciences for guidance in that effort (see Appendix A). In response to her request, the Academy formed the Committee on Declassification of Information for the DOE Environmental Remediation and Related Programs (brief biographical sketches of the members are found in Appendix B).

The Committee held a one-day work session involving officials from the Department and representatives of various affected groups on February 16, 1994. (The work session agenda and list of participants are found in Appendix C; a transcript of the work session is available from the Academy upon request.) The Committee subsequently reviewed an extensive amount of material and held meetings (see Appendix D) to pursue issues raised at the work session, as well as others that arose during the conduct of this study. This report is a result of the Committee's efforts.

During its deliberations, the Committee had substantial assistance from representatives of DOE, especially from the Office of Declassification, from various groups with interests in gaining access to information in the Department's possession, and from interested individuals. Their assistance eased our burden, and we appreciate their help. In addition, we acknowledge the contributions of Gary Stern and William Happer, members of the Committee who withdrew before this report was completed, and the assistance of Jo L. Husbands, director of the Academy's Committee on International Security and Arms Control, for her insightful technical editing of the report.

RICHARD A. MESERVE
Chair

TABLE OF CONTENTS

EXECUTIVE SUMMARY 1
    The Changing Context for Classification Policy, 1
    The Classification System, 3
    Basic Principles for DOE Information Policy, 3
        1. Minimizing the areas that are classified , 3
        2. Shifting the burden of proof , 4
        3. Balancing costs and risks , 4
        4. Enhancing openness and public access , 4
    Taking the Initiative, 5
    Defining the Information Subject to Classification, 6
    Declassifying Documents, 7

INTRODUCTION 11

CHAPTER 1: CONTEXT 15
    A. The Forces for Change, 15
    B. The Competing Interests, 17
    C. Practical and Institutional Obstacles to Change, 20

CHAPTER 2: A DESCRIPTION OF THE CURRENT SYSTEM .. 23
    A. Classification Controls, 23
        1. Two systems, 23
        2. Policy coordination, 26
    B. The Operation of the Classification System, 30
    C. The Freedom of Information Act, 31
    D. Mandatory Review for Declassification, 34
    E. Records Retention and Destruction, 35

CHAPTER 3: BASIC PRINCIPLES AND PRIORITIES
    FOR CHANGE 39
    A. Basic Principles, 39
        1. Minimizing the areas that are classified, 39
        2. Shifting the burden of proof, 40
        3. Balancing costs and risks, 41

4. Enhancing openness and public access, 44
   B. Priorities for Legislative and Regulatory Changes, 47
      1. Amending the Atomic Energy Act (AEA), 47
      2. Using the regulatory process, 50

CHAPTER 4: ISSUES IN CLASSIFICATION POLICY 53
   A. The Case of Nuclear Weapons Information, 54
      1. General policy issues, 54
      2. Transparency vis-à-vis the Russians, 56
   B. The Special Case of UCNI, 59
      1. The original purpose of UCNI, 61
      2. New uses of UCNI, 63

CHAPTER 5: DECLASSIFYING DOCUMENTS 67
   A. Dealing with Existing Documents, 67
      1. Defining the problem, 68
      2. Setting priorities, 69
      3. Making the process work, 72
      4. Increasing effectiveness, 75
      5. Future improvements, 76
   B. Newly Generated Documents, 78

CHAPTER 6: INCENTIVES AND ACCOUNTABILITY 83
   A. Steps to Change the Culture, 83
   B. The Issue of Judicial Review, 85

CHAPTER 7: SUMMARY OF RECOMMENDATIONS 89
   A. Basic Principles, 89
   B. Priorities for Legislative and Regulatory Changes, 90
   C. Issues in Classification Policy, 91
   D. Declassifying Documents, 92
   E. Incentives and Accountability, 93

ABBREVIATIONS USED IN THE REPORT 95

REFERENCES CITED 97

APPENDIX A: CHARGE TO THE COMMITTEE 101

APPENDIX B: BIOGRAPHICAL SKETCHES OF COMMITTEE
    MEMBERS 105

APPENDIX C: WORK SESSION AGENDA AND LIST OF
    PARTICIPANTS 109

APPENDIX D: LIST OF COMMITTEE MEETINGS 113

EXECUTIVE SUMMARY

On December 7, 1993, Secretary of Energy Hazel O'Leary announced an Openness Initiative, the centerpiece of her efforts to make information in areas of concern to the public more accessible. In a news release at that time, the Secretary declared that the goal of this Initiative is "to lift the veil of Cold War secrecy and move the Department of Energy [DOE or the Department] into a new era of government openness." Accordingly, a fundamental reexamination of DOE classification policy and practice is now under way.

Achieving the Secretary's goal of openness and public access to information demands changes both in classification policy and in the process of declassifying and disseminating documents. For example, the Secretary's Initiative was in part a response to public concern about the environmental, safety, and health (ES&H) effects of past DOE activities. DOE has established a policy that all ES&H information is now unclassified in principle, although much of this information is not necessarily available to the public because the documents containing it may also contain other, still-classified information. Before the documents can be made available to the public, they must therefore undergo a painstaking declassification review to guard against the inadvertent release of sensitive information.

Simply finding the relevant material amid DOE's vast collection of classified material is a formidable task. And DOE is losing ground in its declassification efforts -- more new classified documents are being created than old documents are being released.

The Changing Context for Classification Policy

DOE's initiatives take place within a larger, government-wide effort to reexamine classification policy in the wake of the end of the Cold War. U.S. national security policy is no longer directed against the overarching threat of the Soviet Union and its allies. The primary concern of protecting information related to nuclear weapons has shifted to stemming the threat of nuclear proliferation. This complicates some aspects of maintaining the classification system. Protecting information about old nuclear weapons designs or outdated production techniques was formerly considered important but had a lower priority simply because a sophisticated nuclear weapons power like the Soviet Union already had such information. Now, however, protecting such information is essential because the would-be nuclear powers of greatest proliferation concern are less technically sophisticated nations or even terrorist groups, and older (or generally simpler) design and production techniques might better match the capabilities of a potential proliferator.

No foreseeable new nuclear state would pose a threat to the United States and its allies comparable to the threat from the former Soviet bloc. Thus, information that could have helped Soviet bloc war planners, such as the size and composition of fissionable materials inventories or data on most past nuclear weapons activities (but not designs) that might reveal present total capability, is no longer as sensitive as it was once believed to be.

Classification policy must reflect a balance of opposing values. Powerful and compelling reasons continue to exist for protecting genuinely sensitive nuclear weapons information, even though considerable information is already in the public domain. Access to classified information is no longer necessary for a potential proliferator to construct a simple nuclear weapon, but such access could make it significantly easier to build such a device or to make it more effective. The Department would fail in its responsibilities if it did not protect certain design and production information, but the appropriate scope of the information that warrants such careful protection is difficult to define.

In the past, DOE has erred on the side of caution and given the benefit of the doubt to those who argued for classifying many kinds of information. That balance should now be questioned. Secrecy has costs as well as benefits. Secrecy in some areas is inimical to scientific and technical progress. More broadly, the proper functioning of democracy depends on an informed citizenry. Public confidence in government is furthered if independent scrutiny and careful accountability are assured. Making available information that was kept at least partially classified -- ES&H effects of the nuclear weapons complex on the public; the numbers, kinds, and disposition of existing nuclear weapons; and the verifiability of nuclear testing -- would illuminate the debate on important public issues.

Secrecy is also expensive. Creating a classified document imposes a mortgage on the Department to pay for its protection and ultimate review for declassification. DOE estimates that the <u>direct</u> costs of the current classification system are nearly $100 million a year. The <u>indirect</u> costs, in terms of productivity and public accountability, are very difficult to quantify, but are probably several times larger.

The Classification System

The current classification system derives from several separate measures to protect sensitive information. The Atomic Energy Act (AEA) authorizes a system to protect certain nuclear-related information called restricted data (RD). Other, so-called national security information (NSI) is regulated by Executive Order. DOE also exercises control over several other classes of nuclear-related information, such as unclassified controlled nuclear information (UCNI).

The existence of these parallel and somewhat different legal regimes makes the development and application of policy difficult because rules that apply to one category of information do not necessarily apply to another. Declassification of NSI requires agreement among several agencies, and consensus is often elusive. But the divided responsibility for classified information also gives the DOE Secretary an important opportunity to seize the initiative, since under the AEA the Department has independent authority to set policy and declassify documents for many areas of RD.

Basic Principles for DOE Information Policy

1. Minimizing the areas that are classified.

Classification is clearly necessary when uncontrolled release of sensitive information could threaten national security. In a democracy, however, secrecy must be viewed as a necessary evil, to be used sparingly and only with strong justification. DOE should seek to maintain stringent security around sharply defined and narrowly circumscribed areas, but to reduce or eliminate classification around areas of less sensitivity.

2. Shifting the burden of proof.

DOE should be guided by the presumption that information should not be classified unless there is an identifiable reason why the release of the information could damage national security or a reason for concluding that the costs of release outweigh the benefits. The burden of proof should be on those who argue for classification, not on those who propose declassification.

3. Balancing costs and risks.

Information should be classified only if the damage to national security demonstrably outweighs both the public benefit from the disclosure of that information and the costs of attempting to prevent such disclosure. Such a balancing test should be based on an agreed-upon set of criteria for declassification review that has been developed with adequate opportunity for public input. DOE's philosophy should change from the current emphasis on "risk avoidance" to "risk management."

4. Enhancing openness and public access.

The Department's goal should be "open policies openly arrived at." DOE should establish an Information Policy Advisory Board, appointed by the Secretary and composed of experienced outside experts broadly representative of the major stakeholders in DOE's classification policy. The Board would initially provide systematic external input to the fundamental review of classification policy currently under way. Later it could serve a variety of functions, such as making recommendations of priorities for document declassification efforts.

Taking the Initiative

The Secretary of Energy should take advantage of DOE's unique authority to set policy in certain areas of RD. Two important policy changes that do not require approval of other agencies or amending the AEA would be

- Establishing a systematic declassification review of existing documents containing RD, based on priorities reflecting public needs and interests.
- Promulgating a regulation to prohibit abuses of classification under the AEA, comparable to those that now exist for other types of classified information.

In addition, DOE should continue to take the lead in areas where interagency agreement is required and to propose amendments to the AEA to enhance, with the Department of Defense (DOD), the authority to remove some information primarily related to military use of atomic weapons from the category of RD if the agencies determine it can be adequately protected as national security information. This category of information is called formerly restricted data (FRD). Unlike NSI, however, FRD cannot be transferred to any other country except as part of an agreement authorized as part of the Atomic Energy Act. Because this constraint appears needlessly confining, DOE should seek legislative authority to simply transclassify to NSI any RD that no longer warrant special protection as nuclear-related information but still may be sensitive for other military or diplomatic reasons. This would permit elimination of the entire category of FRD.

Where possible, DOE should also develop and adopt any new rules and procedures as regulations using established federal procedures, i.e., notice-and-comment rulemaking. Such regulations would largely replace the current system of DOE orders and would provide a well-understood and accepted mechanism for formal public input into the process. Rulemaking also increases accountability, since decisions are subject to judicial review. Specifically, DOE should promulgate a new regulation concerning classification and declassification of RD and reconsider its existing regulation governing UCNI.

Defining the Information Subject to Classification

DOE's classification policy and its classification guides apply to types and categories of information. Public pressures for greater openness are directed toward four main categories of information: (1) the ES&H effects of DOE activities; (2) the exploitation of classified technologies with potential commercial applications; (3) the historical actions of DOE and its predecessor agencies; and (4) nuclear weapons policy, dismantlement of surplus weapons, and management of the resulting materials. Significant progress has been made in declassifying information in the first two areas, and some progress has been made in the third and fourth. Major issues remain to be addressed, however, especially regarding nuclear weapons information.

The DOE has taken important but only initial steps toward declassification of information that will help inform the public debate about nuclear weapons policy. One example is nuclear weapons tests. A list of all U.S. nuclear tests has been declassified, but only the yields of tests conducted prior to 1962 have been made public. DOE must have agreement from DOD to release the yields of post-1962 tests, and so far consensus has not been achieved.

The United States is currently engaged in negotiations with Russia to persuade it to strengthen controls over the management of fissionable materials from both its weapons and civilian energy programs. As the Committee on International Security and Arms Control of the National Academy of Sciences recommended in its 1994 report, <u>Management and Disposition of Excess Weapons Plutonium</u>, an essential part of this process would be increasing transparency on both sides about the content, location, and management of stockpiles of strategic and other nuclear materials. Russia's secrecy about its nuclear programs is deeply ingrained, and progress to date has been slow. DOE should continue to pursue reciprocal exchanges of information with Russia. Specifically, DOE should explore arrangements in which each party to the exchange retains the right to allow or prevent the public release of the information that it is providing to the other party, so that disagreements about whether information should be publicly released do not obstruct mutually beneficial exchanges. At the same time, however, this objective should not be allowed to delay the release of declassifiable information to the American public.

EXECUTIVE SUMMARY

A category of information that deserves special review is UCNI. When the AEA was amended in 1981 to create UCNI, the primary purpose was to enhance DOE's ability to protect certain information about nuclear facilities, such as floor plans and safeguards, from potential terrorists. Over the years the uses of UCNI have broadened to controlling a wide range of proliferation-sensitive information, such as data with both technical and nuclear weapons-related applications. But the legislative base for UCNI was never updated to reflect this expanded scope. The Committee has received no persuasive justification for the continuation of UCNI as a special category and recommends that DOE undertake a thorough reexamination of whether there is a continuing need for these controls. If DOE concludes that information now encompassed by UCNI should continue to be protected under this scheme, it should prepare a clear and thorough justification for the proposed uses of UCNI and a comparison of alternative approaches to achieve the same objectives.

When the fundamental review of classification policy is completed, DOE should indicate publicly which areas of information it believes no longer require protection as RD, even if it may not have interagency agreement to declassify all these areas. Its draft report on classification policy, "Public Guidelines to Department of Energy Classification of Information," should be released promptly and made readily accessible.

Declassifying Documents

Narrowing the information subject to classification is not in itself sufficient to achieve a policy of openness. DOE must find a means for reviewing documents to determine whether they contain only unclassified information and then release them to an often skeptical public. It also needs to find ways to minimize the creation of new classified documents and to reduce the effort required at some future date in declassifying the documents and their copies and derivatives.

DOE has made efforts to declassify and publicly release many of the documents that have accumulated over the last 50 years. A massive task confronts the Department. Using DOE's current estimate that it holds approximately 280 million classified pages -- and we consider this estimate very uncertain -- then simply reviewing the current holdings using the standard methods would require almost 9,000 person-years of effort. Better, faster, and less costly methods must be found.

At present DOE is only beginning to assess the magnitude of the declassification task it faces. Since resources will be limited, priorities must be set to guide the effort. The process must be demand-driven, addressing the areas of greatest public concern first, particularly during the interim period while DOE's declassification policy is undergoing fundamental review and change. Once new DOE policy is established (with appropriate stakeholder involvement), a more regular process that gives the public a significant voice in setting priorities should be created. Strong local and regional inputs are essential.

The Department needs to take other steps as well. The number of reviewers must be increased. DOE's plans for declassification should include planning for the preparation of a record index (with unclassified title, author, date, and document number) to be made available to the public and updated periodically.

The Department needs to develop and evaluate more cost-effective declassification methods. Bulk declassification is not promising, since the methods suggested to date pose too great a risk that sensitive information could be released unwittingly. It may be possible, however, to establish methods for quickly screening large numbers of documents, sorted according to those most likely to contain sensitive information. Those least likely to contain such information could be given priority and could be subjected to less rigorous scrutiny. DOE also needs to investigate new methods, such as artificial intelligence, that offer promise of reducing the burden in the future.

To minimize the generation of new classified documents, DOE needs to institute a number of new procedures. Classified or otherwise controlled information should be included in documents only if absolutely necessary. Strict guidance should be provided for use of derivative classification, which occurs when new documents make reference to material in classified documents, and must therefore be themselves classified. Portion marking of those areas containing classified information should be required, as well as segregating the classified portions whenever possible. For documents containing information of significant interest to the public, preparing unclassified versions could help where segregation is not practical. Documents should be coded and indexed so they can be easily tracked, identified, and reviewed for declassification when guides change.

Changes in policy and formal procedures will have only limited impact if significant effort is not devoted to changing the Department's traditional modes of operation, which place a strong emphasis on the costs of openness and little emphasis on the costs of secrecy in a democratic society. In the current system, incentives favor overclassification. The potentially irreversible consequences of a mistaken decision to declassify reinforce the tendency to overclassify.

Measures to provide incentives for changing behavior, such as including measures of openness in performance evaluations for agency personnel and contractors, should be instituted. At the level of policy, DOE should require a substantive justification in terms of explicit criteria for keeping an area classified whenever it is subject to declassification review. Without such sustained effort, the inertia of traditional practice and the sheer size of the challenges that must be overcome to achieve meaningful change will needlessly delay reform efforts.

Secretary O'Leary and DOE have undertaken important initiatives to achieve greater public access to information and greater departmental accountability for the information DOE controls. The Committee on Declassification of Information for the DOE Environmental Remediation and Related Programs commends these efforts. The Openness Initiative is an essential part of the U.S. government's reevaluation of the classification system in the aftermath of the Cold War. The availability of more information about past and current policies should help to restore public confidence in institutions and to foster a more informed public debate on essential policy choices.

INTRODUCTION

Secretary of Energy Hazel O'Leary has declared that her goal is to make information more accessible wherever possible and appropriate. She has launched a program to disclose previously classified information in such areas and to modify classification practices. The centerpiece of this program is the Openness Initiative, announced on December 7, 1993, "to lift the veil of Cold War secrecy and move the Department of Energy (DOE) into a new era of government openness."[1]

The problems of classification represent a complex mix of policy and process issues. Defining the appropriate boundary for classification requires a balancing of the cost and benefits of secrecy and of openness. The necessary calculus is complicated and involves significant elements of judgment.

Some of the difficult issues are one step removed from the classification problem itself. For example, a major reason for the Secretary's Openness Initiative was the intense public interest in the environmental consequences of activities in the DOE weapons complex and in the studies of the health effects of radiation exposures. All information of this sort -- about environmental, safety, and health (ES&H) effects of DOE programs; biological effects of radiation; and research and development concerning medical, biological, health and safety, and environmental studies -- is now unclassified in principle.[2] But the information is not necessarily available to the public because many of the documents containing such information may also contain material that is still classified. The documents must therefore undergo a painstaking process of declassification review to determine if they, or the information they contain, can be publicly released. At present there is no comprehensive program for declassification or review of classified documents required by law, except for those considered in response to Freedom of Information Act (FOIA) requests.

---

[1] U.S. Department of Energy Office of Press Secretary, 1993, p. 1.

[2] U.S. Department of Energy Office of Classification, 1994a, p. 40.

Identifying the relevant materials in the huge collection of documents under the Department's control is a formidable challenge. Currently DOE estimates that it holds about 280 million classified pages (including copies),[3] an estimate up by a factor of 10 over the one given in the Secretary's December 1993 press conference. And the Committee has been told that DOE is losing ground in its declassification efforts -- more new classified documents are being generated than old documents are being declassified.[4] Thus, in addition to policy questions, the Openness Initiative must address difficult issues of document examination, document control, and public communications. Achieving the Secretary's goal of openness and public access to information potentially affects large areas of DOE policy and operations.

Secretary O'Leary does not lack sources of expert advice on these problems. Of particular importance, the Administration has issued Exec. Order No. 12,958 -- the order that governs the classification, declassification, and control of certain classified information throughout the federal government.[5] The Joint Security Commission, at the request of the Secretary of Defense and the Director of the Central Intelligence Agency, has issued a comprehensive report advocating a new approach to security issues.[6] The Congress has established a Commission on Protecting and Reducing Government Secrecy to conduct a thorough review of security issues.[7] To the extent possible, DOE's policy needs to

---

[3] Estimate from the U.S. Department of Energy Office of Information Resources Management Policy, Plans, and Oversight, February 1995.

[4] P.R. Laplante, DOE Office of Declassification, personal communication, 1994.

[5] This Executive Order, replacing Exec. Order No. 12,356, was released on April 17, 1995.

[6] Joint Security Commission, 1994.

[7] The declared purpose of the Commission is "(1) to examine the implications of the extensive classification of information and to make recommendations to reduce the volume of information classified and thereby to strengthen the protection of legitimately classified information; and (2) to examine and make recommendations concerning current procedures relating to the granting of
(continued...)

be in harmony with the overall government strategy for control of classified information.

DOE is also engaged in its own evaluations. The Department began a broad review of classification policies and procedures in 1990, leading to publication of a study of classification issues in 1992.[8] A key recommendation of that study was that DOE should "[c]onduct a comprehensive, fundamental review of all nuclear weapon-related information to determine what should be classified under present conditions, with the objective of removing from classification all information that no longer warrants such protection."[9] Secretary O'Leary included such a review as part of her Openness Initiative. According to DOE, this effort will evaluate "all existing classification policies and related technical guidance to determine what still makes sense to classify under current world conditions." The review will

- Issue revised broad policy criteria based on current world conditions and domestic objectives.
- Apply revised broad criteria in the re-evaluation of detailed technical policy and guidance, and update and issue classification guides at DOE headquarters and in the field to reflect revised policies.
- Develop procedures for continued, periodic re-evaluations of policy to keep pace with future world conditions and domestic objectives.
- Develop enhanced automation techniques to support the declassification process by accelerating guidance update and issuance.[10]

(...continued)
security clearances." Protection and Reduction of Government Secrecy Act, Pub. L. No. 103-236, § 903, 108 Stat. 525, 526 (1994).

[8] Meridian Corporation, 1992.

[9] Meridian Corporation, 1992, p. 7.

[10] Keliher, 1994.

The DOE Fundamental Classification Policy Review held its first meeting on March 16, 1995.

The Committee therefore finds itself in the difficult position of having to comment on a complex subject while it is under far-reaching examination and change by others. DOE has launched efforts to declassify information, while at the same time the overall framework for classification is under study in a variety of fora. The subject of our inquiry has thus evolved and mutated during the course of our scrutiny. We have sought, therefore, to provide an overview that will offer a foundation and framework for evaluating the diverse policy advice that is now or will shortly be available on classified matters.

Chapters 1 and 2 of this report discuss the context for the classification system and describe its current operation. Chapter 3 lays out the fundamental principles that guided the Committee's assessment and suggests several basic legislative and regulatory changes. Chapter 4 addresses the complex issues of information policy, and Chapter 5 assesses ways to improve the declassification process. Chapter 6 examines changes in incentives and accountability that will be necessary to significantly alter the current culture of DOE operations. Finally, Chapter 7 presents a summary of the recommendations that appear in Chapters 3 through 6.

CHAPTER 1

CONTEXT

An assessment of the Department of Energy's (DOE) classification system must be guided in part by awareness of the forces for change, the competing interests, and the practical challenges confronting the Department.

A. The Forces for Change

The present classification system was established during World War II and evolved during the Cold War. Profound changes in international conditions affecting the role of the Department have put new pressures on the system.

With the end of the Cold War, U.S. national security policy is no longer directed against the overarching threat of the Soviet Union and its allies. The nation's primary concern relating to the secrecy of nuclear weapons information has shifted from bilateral nuclear confrontation with the Soviet bloc to a growing concern with nuclear proliferation.

The new international conditions may make some aspects of maintaining a classification system more difficult. Because the former Soviet Union had a sophisticated nuclear weapons capability, extreme measures were not necessary to prevent its access to old designs or outdated production techniques for the simple reason that Soviet technologists already possessed such information. The new priority given to nonproliferation requires keeping information from nations or parties that are less technically sophisticated than the former Soviet Union. This creates a continuing need to protect older (or generally simpler) designs and production techniques, because the older approaches might be more commensurate with the capabilities of a potential proliferator. A workable World War II-era bomb design, while perhaps having a lower and less reliable yield than a more modern design, is nonetheless extremely dangerous. In addition, other categories of information -- such as data, planning documents, and instructions pertaining to ongoing negotiations with foreign powers; technical design items relating to nuclear delivery and defensive systems; and safeguards information -- still merit security protection.

On the other hand, the elimination of the threat of nuclear war with the Soviet Union means that certain information that could have benefitted Soviet bloc war planners no longer needs such careful protection. No foreseeable proliferator poses a comparable nuclear threat to the U.S. and its allies. Our concern with potential proliferators (and their concern with us) will be independent of the specifics of the vastly superior nuclear stockpiles of Russia and the United States. Thus, for example, information about the size and composition of fissionable material inventories is no longer as sensitive as it once was believed to be. Similarly, data concerning past nuclear weapons activities that would reveal our present total capability are no longer sensitive because our capability greatly exceeds that of any potential adversary.

Changing world circumstances have also had indirect effects on the Department, altering the balance of costs and benefits that must guide the declassification system. The end of the Cold War, current federal budget realities, and other public forces have changed DOE's mission. DOE is now working to address the environmental legacy of its past activities. That process includes the remediation of contaminated facilities in the weapons complex; the evaluation of the environmental, safety, and health (ES&H) consequences of the facilities; and the public acknowledgment of past activities, some of which were wholly at odds with current requirements. While DOE's stewardship for the nuclear weapons complex remains an important responsibility, the Secretary has correctly decided that the Department cannot restore the confidence of the public that it is forthrightly confronting its past without a policy of greater public access to information.

There are also strong external pressures on DOE for increasing access to information. Public interest groups and citizens continue to believe that information necessary for discussions of safety and health is inaccessible because of classification or other information controls. Moreover, according to the U.S. Congress Office of Technology Assessment, "[v]irtually all public interest groups concerned with nuclear weapons issues...share common concerns about...the public's access to relevant information..."[1] States affected by DOE weapons complex sites are seeking more rapid declassification of documents that could contain important information relevant to the cleanup of those sites.[2] DOE's changing mission and the public pressures on it thus reinforce the modifications in the declassification system's focus already arising from the end of the Cold War.

B. The Competing Interests

DOE's classification policy must reflect a balance of opposing values. On the one hand, there are powerful and compelling justifications for protecting certain information from public disclosure. A core mission of DOE has been the development of nuclear weapons. Although considerable information concerning the design and construction of nuclear weapons is in the public domain, access to a portion of it must be severely restricted because it could be exploited by a terrorist or by a state seeking to develop or improve nuclear arsenals. DOE would fail in its responsibilities if it did not protect such information. An effective classification system to define and control such information serves the interests of this country as well as those of other nations. Indeed, limiting access to some types of nuclear information is essential for the welfare of all mankind.

It is difficult to define appropriately the scope of the information that should be subject to classification. All agree that it should encompass information that is not in the public domain and that is essential to the design and construction of a nuclear weapon. But how widely should the classification net extend? Should it include the numbers, kinds, or disposition of existing nuclear weapons? operational plans for employing nuclear weapons? data relating to the existence, nature, and results of nuclear tests? information about measures to prevent nuclear accidents? information on foreign nuclear programs? information on dual-use technologies that could be applied both to nuclear weapons and to civilian products or processes?

---

[1] Office of Technology Assessment, 1993, p. 113.

[2] Oregon Department of Energy, 1994; Morin, 1994.

In defining appropriate boundaries for the classification system, the costs and benefits of secrecy must be weighed with the costs and benefits of openness. Although the benefits of secrecy in some areas may be paramount, the full consideration of costs and benefits will certainly justify openness in others. The benefits of disclosure and dissemination are significant.

Fundamentally, the proper functioning of democracy depends on an informed citizenry. Making information available that in part has been kept classified -- ES&H effects of the weapons complex on the public; the numbers, kinds, or disposition of existing nuclear weapons; and the verifiability of nuclear testing -- would illuminate the debate on and scrutiny of important public issues.

Public confidence in governmental institutions is furthered if independent scrutiny and careful accountability are assured. Governmental secrecy undermines confidence in public institutions.[3] Indeed, classification can be a "refuge for scoundrels" because there is danger that information might be classified simply to prevent disclosure of imprudent, unethical, or illegal actions.

Secrecy in some areas is also inimical to scientific and technical progress. The source of valuable ideas cannot be predetermined, and therefore secrecy, in limiting the number of persons who can have access to specific items of information, can have a serious impact on progress and creativity. Open communication on technical issues not only allows the expansion of scientific and technical knowledge, but also expedites opportunities for commercial applications.

The direct financial costs associated with maintenance of the classification system are substantial. Classified information must be protected until public release, and personnel allowed access to the information must be cleared for this access. Creating a classified document thus imposes a mortgage on DOE to pay for protection of that document. DOE estimates that the <u>direct</u> costs of its classification system (e.g., costs of physical and electronic security measures, classification management, and personnel security) are almost $100 million a year. This is a small but significant fraction of the nearly $2.3 billion spent government-wide on

---

[3] Office of Technology Assessment, 1993, p. 109-123.

classification-related activities, most of it by the Department of Defense.[4] The indirect costs, in terms of productivity and public accountability, are very difficult to quantify, but may be many times larger.

In defining the appropriate balance between secrecy and openness, one must also consider the effectiveness of measures designed to assure secrecy. One example shows the effectiveness of past classification of restricted nuclear information. The Congressional Research Service reported, and sources from DOE laboratories confirmed, that while the United Nations inspectors in Iraq discovered an extensive nuclear-weapons enterprise and noted egregious violations of export regulations, they found no evidence of significant Iraqi knowledge of classified information at the level of restricted data (RD).[5] This is gratifying information, but it provides little guidance for evaluating the overall effectiveness of security measures.

The chain of protection surrounding classified information is no stronger than its weakest link. Although the Iraqis do not appear to have gained access to U.S. classified information relating to nuclear weapons, numerous instances of outright criminality by people with access to classified information can be cited. These spy cases make the limits of the classification system evident. While the administration of classification controls is not within the purview of this Committee, we cannot assume that classification serves as a long-term, assured system for denying a potential adversary access to information.

Moreover, a great deal of nuclear information is already publicly available. The fundamental principles of nuclear weapons construction are well understood and exist in the open literature. Basic physical and chemical data on materials, including fissionable materials at high temperatures and high pressures, are largely unclassified. The process for separating plutonium from spent fuel is covered in unclassified treatises on the chemistry of such methods. Access to classified information is not necessary for a potential proliferator to construct a nuclear weapon. Such access makes it easier to construct a nuclear device or to construct a more effective weapon, but the fundamental information needed for making a

---

[4] Office of Management and Budget, 1994.

[5] Zimmerman, 1993.

very destructive device is widely available. The classification system alone cannot keep the nuclear monster in its cage.

In sum, defining the boundaries of the classification system requires consideration of a variety of competing factors. Achieving an appropriate balance is all the more difficult because some of the factors cannot be quantified or evaluated with confidence.

C. Practical and Institutional Obstacles to Change

DOE's efforts to achieve greater public access to information face a number of significant practical and institutional obstacles. The first, as mentioned, is the huge and still growing collection of classified documents held by the Department and its contractors. Many of the documents may have been classified under guidance that has since been revised and hence are no doubt appropriate for declassification and public release without any change in Departmental policy. But the process of declassifying such documents is time-consuming and expensive.

Another obstacle is the fact that any change in Department-wide policy may have only limited and delayed practical impact at the level at which classification actions are taken. All government programs have a certain measure of "momentum," by which we mean that changes in actual behavior occur slowly and only with the application of significant public, political, or other force. The usual pattern of incremental, grudging acceptance of change can be expected in connection with changes in classification policy. From its beginnings, DOE and its predecessor agencies maintained a culture in which, for understandable reasons, protecting information was dominant. That environment has had an impact on personnel. The prestige, status, and perceived self-worth of staff are related to access to information that is denied to others.[6] Changing such practices, deeply embedded in DOE's reward and status structure, is not a trivial undertaking, but without such changes, simply altering policies may have limited effect.

Third, there is a "ratchet effect" in the handling of classified information. Although policymakers may seek to place the burden on those who would classify information, the reality may be different.

---

[6] Gusterson, 1992.

Information, once released to the public, is difficult or impossible to recapture.[7] This provides practical pressure to maintain classification, because the perceived consequences of an incorrect decision are unequal -- that is, an incorrect decision to classify information can be corrected, but an incorrect decision to release information may be irreversible. The natural response to such a situation is to be very cautious about declassification. The avoidance of overclassification requires a process that minimizes errors in connection with either classification or declassification decisions. But setting up a procedure to give such assurances serves only to increase the cost and personnel demands associated with the effort.

In addition, DOE does not have autonomy regarding all the types of information it classifies and declassifies. As discussed in the next chapter, DOE must coordinate policies and procedures for certain types of nuclear weapons information with other departments, particularly DOD. Achieving interagency consensus on greater openness and public access to information can be difficult.

Finally, policies regarding classified information are now being reviewed throughout the federal government. Although DOE can act alone with respect to some kinds of classified information, its program should be developed in a fashion that harmonizes with the overall approach.

---

[7] In addition to the practical difficulties of recapturing released information, DOE interprets the Atomic Energy Act as prohibiting reclassification of information once an area of Restricted Data (RD) has been declassified (U.S. Department of Energy, 1994, p. 20).

CHAPTER 2

A DESCRIPTION OF THE CURRENT SYSTEM

A complex matrix of statutes, regulations, and procedures governs the control of classified information, public access to governmental information, and the maintenance of governmental records. This matrix defines the context for the Department of Energy (DOE) classification system. The following overview describes the legal foundation for the system.

A. Classification Controls

1. Two systems

Documents are classified for national security reasons under two different systems. Most of the classified information held by the federal government is classified, pursuant to Exec. Order No. 12,958, as national security information (NSI). An affirmative act by a government official is required to designate information as NSI. Categories of information that are eligible for classification as NSI under the current executive order include (a) military plans, weapons, or operations; (b) foreign government information; (c) intelligence activities (including special activities), intelligence sources or methods, or cryptology; (d) foreign relations or foreign activities of the United States, including confidential sources; (e) scientific, technological, or economic matters relating to the national security; (f) United States Government programs for safeguarding nuclear materials or facilities; or (g) vulnerabilities or capabilities of systems, installations, projects or plans relating to the national security.[1]

---

[1] Exec. Order No. 12,958, § 1.5.

Under the second system, certain nuclear-related information, called restricted data (RD), is classified according to a system created by the Atomic Energy Act of 1954 (AEA).[2] The AEA provides:

> The term "Restricted Data" means all data concerning (1) design, manufacture, or utilization of atomic weapons; (2) the production of special nuclear material; or (3) the use of special nuclear material in the production of energy, but shall not include data declassified or removed from the Restricted Data category pursuant to section 2162 of this title.[3]

The scope of the definition is broad and is rendered even more elastic by expansive definitions of "design" and of "research and development."[4] Unlike NSI, RD is interpreted by DOE as "born classified" -- that is, to be considered a protected secret upon coming into existence without any affirmative act or decision by an official or, indeed, any involvement by government at all.[5] The AEA authorizes sealing off an entire area of scientific and engineering knowledge from public scrutiny.

---

[2] 42 U.S.C. §§ 2011 et seq.

[3] 42 U.S.C. § 2014(y).

[4] The AEA provides that "[t]he term 'design' means (1) specifications, plans, drawings, blueprints, and other items of the like nature; (2) the information contained therein; or (3) the research and development data pertinent to the information contained therein" (42 U.S.C. § 2014(i)).

"Research and development" are defined as "(1) theoretical analysis, exploration, or experimentation; or (2) the extension of investigative findings and theories of a scientific or technical nature into practical application for experimental and demonstration purposes, including the experimental production and testing of models, devices, equipment, materials, and processes" (42 U.S.C. § 2014(x)). Neither definition is readily confined.

[5] Hewlett, 1981; Green, 1981.

The AEA has provisions authorizing declassification of information falling within the scope of the definition.[6] Over the years, RD relating to many once-classified areas has been declassified, largely in order to facilitate commercial applications.[7] As a result, information relating to civil power reactors and nuclear fuel reprocessing is not classified. The remaining areas of national defense-related nuclear information that contain RD pertain to (1) nuclear weapon design; (2) nuclear material and nuclear weapon production; (3) certain theoretical aspects of inertial confinement fusion; (4) military reactors (production and submarine reactors); (5) isotope separation; and (6) directed nuclear energy systems.[8]

Only those with security clearances are given access to either NSI or RD.[9] The levels of classification in the two regimes are identical -- Top Secret, Secret, and Confidential -- and only individuals with the requisite level of clearance can obtain access. Detailed requirements for the safeguarding and control of classified information are keyed to the level and type of classification.

Having a clearance, however, is not sufficient to gain access to classified information. An individual must also be shown to have a "need to know" the information in question. There is a formal system for controlling access to certain areas of information on a need-to-know basis. This additional layer of categories and controls adds to the complexity of the system.[10]

---

[6] 42 U.S.C. § 2162.

[7] Some of the declassified information is still subject to control as unclassified controlled nuclear information (42 U.S.C. § 2166).

[8] Meridian Corporation, 1992, p. 23.

[9] Access to NSI is generally limited to those for whom access is required "in order to perform or assist in a lawful and authorized governmental function" [Exec. Order No. 12,958, § 4.1(c)]. With appropriate protection against general disclosure, the access limitations may be waived for historical researchers [Exec. Order No. 12,958, § 4.5(a)(1)].

[10] Meridian Corporation, 1992, p. 87.

This report deals primarily with information that is now or was once classified as RD. Because of DOE's special mission relating to development, testing, and production of nuclear weapons, many DOE facilities, including the complex of national laboratories and other nuclear weapons-related facilities, have extensive files dating from the beginning of the atomic age that contain RD.

2. Policy coordination

The existence of different legal regimes covering classified information has important practical implications for DOE. On the one hand, the existence of parallel (and somewhat different) legal regimes makes the development and application of policy difficult because rules that apply to one category of information do not necessarily apply to the other. Declassification of NSI is not subject to the Secretary's control, but is possible only by agreement with other agencies under terms defined by Exec. Order No. 12,958. On the other hand, the Secretary has primary authority to set policy for declassification of most RD.

The principal exception to the Secretary's control over RD concerns information related primarily to the utilization of nuclear weapons. The AEA provides that DOE and the Department of Defense (DOD) must jointly decide whether information in this category can be publicly released without "unreasonable risk to the common defense and security."[11] DOE also shares with DOD the authority to remove some information primarily related to military use of atomic weapons from the category of RD if the agencies determine it can be adequately protected as national security information.[12] Such information, called formerly restricted data (FRD), cannot be transferred to any other country while it remains classified as NSI except as part of an agreement authorized by the AEA.

Some of the information sought by those interested in participating in policy debates about nuclear weapons, nonproliferation, dismantlement, and materials disposition fall into the category of NSI, for which

---

[11] 42 U.S.C. § 2162.

[12] 42 U.S.C. § 2162.

concurrence on declassification from other agencies may be necessary (see Table 1). The DOD, Department of State, Arms Control and Disarmament Agency, Central Intelligence Agency, National Security Council, and Joint Chiefs of Staff all have interests in how such data are handled. DOE's ability to declassify information relevant to policy debates about nuclear weapons is thus dependent in part on the willingness of other agencies with different priorities to cooperate in a timely manner.

Table 1

Agency Responsibility/Concurrence to Declassify Various Categories of Nuclear Weapons Information

Information Category	Responsibility for Action	Concurrence Needed
Nuclear weapons in active use	DOD	DOE
Inactive stockpile	DOD	DOE
Warhead modifications and retirements	DOD	None
Weapons to be disassembled, already disassembled, and disassembly rates	DOD	DOE
Fissile material stockpile		
In weapons	DOE	None
Available for use in weapons	DOE	None
Unavailable for weapons	DOE	None
Historical data on fissile material production	DOE	None
Weapon test name associated with		
Yield	DOE	DOD
Mark or W number	DOE	DOD
Device nickname	DOE	None

SOURCE: Information provided by R. Lyons, DOE Office of Declassification, September 7, 1994.

The National Security Council has designated DOE as the lead agency for an interagency process to formulate declassification policies and specific actions on nuclear weapons-related matters. Disagreements are resolved by the Nuclear Weapons Council, which includes the Under Secretary of DOE and the Deputy Secretary of Defense. This process has led to the recent declassification of some weapons-related information, but has revealed significant differences between the agencies. For example, the 1992 DOE-sponsored study of classification policy recommended that data concerning the occurrence of all nuclear tests and their yields should be declassified.[13] While DOE has now declassified the total list of past nuclear test explosions, no agreement has been reached with DOD on the declassification of yields of post-1962 tests. Similarly, the declassification of information concerning verification activities related to nuclear tests and production activities of fissionable material has been impeded by the need for interagency coordination.

RD and NSI are not the only classes of information that are subject to control by DOE. The Department also exercises control over unclassified information in a number of areas. These include unclassified controlled nuclear information (UCNI), naval nuclear propulsion information (NNPI),[14] export control information (ECI),[15] and official use

---

[13] Meridian Corporation, 1992, p. 84.

[14] Order DOE 5630.8A (U.S. Department of Energy Office of Nuclear Reactors, 1990) defines any information with an identifiable association with naval nuclear propulsion as NNPI. NNPI may be either classified or unclassified information. A small subset of NNPI -- specifically, all unclassified information related to the reactor plants of naval nuclear propulsion -- is also classified as UCNI. The order allows for dissemination of NNPI only on a highly restrictive need-to-know basis, and provides special procedures for marking, handling, and access. These requirements may be in addition to those imposed by the UCNI rules. However, the order contains no sanctions other than those otherwise available for protection of NSI, RD, or UCNI.

[15] The DOE Office of General Counsel holds that DOE does not have power to control ECI generated within the DOE complex, although such information would be subject to control if generated privately. DOE has developed voluntary guidelines for control by the laboratories, but they are controversial (Meridian (continued...)

only information.[16] This report will focus on only one of these categories -- UCNI -- which is the most far-reaching. As will be discussed, UCNI encompasses unclassified information that relates to the physical security of nuclear facilities and, as currently interpreted, other information relating to technology that could be useful to a potential proliferator.

Finally, the United States has certain obligations to protect information as a result of its international treaty commitments. Article I of the Treaty on the Non-Proliferation of Nuclear Weapons provides that

> Each nuclear-weapon State Party to the Treaty undertakes not to transfer to any recipient whatsoever nuclear weapons or other nuclear explosive devices or control over such weapons or explosive devices directly, or indirectly; and not in any way to assist, encourage, or induce any non-nuclear weapon State to manufacture or otherwise acquire nuclear weapons or other nuclear explosive devices, or control over such weapons or explosive devices.[17]

The nuclear weapon state parties to the treaty -- the United States, Russia, China, France, and the United Kingdom -- are implicitly prohibited from disseminating information or data concerning the design and construction of nuclear weapons. The treaty obligation thus bears on declassification actions relating to nuclear weapons, with the same intent as domestic U.S. legislation designed to protect nuclear information.

(...continued)
  Corporation, 1992, p. 67-68).

[16] This is sensitive information, which might be exempt from the Freedom of Information Act (FOIA) under certain circumstances. Rules exist for several DOE programs, but not a DOE-wide approach. The Classification Policy Study (Meridian Corporation, 1992, p. 67-68) recommends that one should be adopted.

[17] Treaty on the Non-Proliferation of Nuclear Weapons, 21 U.S.T. 483, March 5, 1970.

B. The Operation of the Classification System

Approximately 5,000 individuals in the Department and its contractors have formal authority to determine whether a document should be classified and, if so, at what level. These decisions are governed by a detailed Classification Guide that sets out criteria as to whether a given fact is classified and, if so, its level of classification. Some, but not all, of the volumes in the Classification Guide are themselves classified.

Some DOE facilities have developed derivative classification guidance that, although based on the Classification Guide, provides information on classification that is particularly relevant to the activities of each facility. There are some 880 detailed classification guides at DOE headquarters and field offices.[18] We understand that these derivative guides are reviewed by the Office of Declassification to ensure consistency with overall Department classification policy.

Because the scope of the information subject to classification has changed over the years and the markings on existing documents are not automatically or periodically revised, the current inventory of classified documents may contain documents that were properly classified at the time of creation, but that are not properly classified under current policy. All such documents are treated as classified until an affirmative decision is made to modify their classification status. There is no comprehensive program for declassification or review of classified documents required by law, except for those considered in response to Freedom of Information Act (FOIA) requests.

---

[18] In practice, the experts in a given field have a working knowledge of the aspects of their work that are classified. Such individuals no doubt mark a document as classified without detailed review of the Classification Guide or the derivative guides.

C. The Freedom of Information Act

The principal means for public access to public records is through FOIA.[19] FOIA specifies, as a basic right, public access to the records of all federal agencies, but provides for categories of information that are exempt from disclosure. FOIA lists nine exemptions, but only two - one protecting NSI (Item 1 in footnote 20) and another protecting RD and UCNI (Item 3 in footnote 20) - are relevant to the work of the Committee.[20]

When "any person" makes a request for records that "(A) reasonably describes such records and (B) is made in accordance with published rules," the agency must make the records "promptly available."[21] DOE has developed its own procedures for implementing FOIA. Requests must be in writing and must reasonably describe the record sought.[22] Categorical requests -- requests for all information in a "reasonably specific and well-defined category" -- are permitted and handled similarly to other requests.[23] When a FOIA request involves records or information that originated in another agency, DOE must refer the request to the originating agency or get approval from the originating agency to make the

---

[19] 5 U.S.C. § 552.

[20] The exemptions exist for (1) secret national security information, (2) internal personnel rules and practices of the agency, (3) matters specifically exempted from disclosure by statute, (4) trade secrets, (5) interagency or intra-agency memoranda that would be protected in litigation involving the agency, (6) personnel and medical files affecting personal privacy, (7) records or information compiled for law enforcement purposes, (8) agency reports concerning regulation of financial institutions, and (9) matters concerning geological information [5 U.S.C. §§ 552(b)(1-9)(1988)]. The Supreme Court has stated that "[t]hese exemptions are specifically made exclusive '... and must be narrowly construed'" [Department of the Air Force v. Rose, 425 U.S. 352, 361 (1976), quoting EPA v. Mink, 410 U.S. 73, 79 (1973)].

[21] 5 U.S.C. § 552(a)(3).

[22] 10 C.F.R. § 1004.4(b).

[23] 10 C.F.R. § 1004.4(c).

decision. If the latter, DOE must coordinate its response with that agency.[24]

In response to a proper request, the Freedom of Information Officer sends a copy of the request to the Authorizing Official (the official having custody or responsibility for the requested records).[25] DOE treats requests for classified records similarly to requests for nonclassified information, but requires that the Director of Classification (now Declassification) receive notification of any request for classified information, advise the Authorizing Official processing the request, and concur in the determination.[26] When a request, or part of a request, is denied because the withheld information is classified, the report of denial must list the name of the Director of Classification as the Denying Official for the withheld classified matter.

Each notified Authorizing Official must prepare a written response within 10 working days of receipt of a request, except in the case of "unusual circumstances," which may include the need to search for and retrieve records from other DOE offices, the need to examine a large volume of requested materials, or the need to consult with another agency that has "substantial interest in the determination of the request."[27] The agency and the requester may agree to extend the period for initial DOE response to an FOIA request, but if DOE has not made a decision on a request by the end of the 10-day period or the extended period, the requester may file for review of his or her claim in federal district court.[28]

---

[24] 10 C.F.R. § 1004.4(f).

[25] 10 C.F.R. §§ 1004.2, 1004.5(a)-(c). The regulations designate Freedom of Information Officers at eighteen regional DOE offices to oversee the processing of FOIA requests.

[26] 10 C.F.R. §§ 1004.6(c)-(d).

[27] 10 C.F.R. § 1004.5(d).

[28] 10 C.F.R. §§ 1004.5(d)(4)-(5).

When an FOIA request is approved, the records "will be made available promptly."[29] When a request is denied, the requester must be provided with a written explanation, including reasons for denial, the persons responsible for the denial, information about whether the requested information contains nonexempt material that can be segregated, and notice of the availability of appeal challenging either the adequacy of the search or the decision to deny the request.[30] A person whose FOIA request has been denied may appeal to the Office of Hearings and Appeals (OHA) within 30 calendar days of notice of denial.[31]

As of June 1994, DOE had 812 open FOIA requests at DOE headquarters and another 450 requests in the field. Of these, 82 had been initiated in the 1980s; at her press conference on June 27, 1994, the Secretary committed to closing these requests by the end of that year. By the end of 1994, only two requests remained open. (It should be noted that closure of an FOIA request does not necessarily mean that the document requested was found and supplied.) During the last seven months of 1994, DOE headquarters closed 520 requests compared with 236 over the same period in 1993. Perhaps as a result of the Openness Initiative, FOIA requests more than doubled from July to December 1994, so that in mid-February 1995 there were 727 FOIA requests pending at headquarters.[32]

---

[29] 10 C.F.R. § 1004.7(a).

[30] 10 C.F.R. § 1004.7(b).

[31] 10 C.F.R. § 1004.8(a). The appeal must be in writing and must discuss the legal bases for the appeal [10 C.F.R. § 1004.8(b)]. The OHA must act on the appeal within 20 working days, except in "unusual circumstances" similar to those outlined above [10 C.F.R. § 1004.8(b)]. Failure to issue a decision within the statutory period constitutes an exhaustion of administrative remedies, allowing the requester to seek review in a district court (10 C.F.R. § 1004).

[32] R. Lyons, DOE Office of Declassification, personal communication, February 22, 1995.

D. Mandatory Review for Declassification

Section 3.6 of Exec. Order No. 12,958 requires that, in response to a proper request, an agency shall conduct a mandatory review for declassification of NSI. (The provision does not apply to RD, and there is no comparable provision in statute or regulation for an externally initiated mandatory declassification review of RD.) The request must be made by a U.S. citizen or a permanent resident alien, a federal agency, or a state or local government.[33] The request must describe the subject matter "with sufficient specificity to enable the agency to locate it with a reasonable amount of effort."[34] Exec. Order No. 12,958 additionally directs agencies to develop procedures to conduct mandatory declassification review, including an appeals process, and requires agencies to declassify items of NSI that no longer need to be kept secret.[35]

DOE had issued separate regulations implementing the previous executive order (Exec. Order No. 12,356).[36] These guidelines give DOE's Assistant Secretary for Defense Programs authority to make final determinations on appeals of denied requests for NSI under the Mandatory Review for Declassification provision.[37] The regulations require that DOE, in answering a valid request for declassification review, coordinate its review with any other agency to which the NSI is relevant.[38]

---

[33] This category of eligible requesters is narrower than that in the FOIA, which allows "any person" to make a request (Adler, 1993).

[34] Exec. Order No. 12,958, § 3.6(a)(1).

[35] Exec. Order No. 12,958, §§ 3.6(c)-(d), 5.4.

[36] 10 C.F.R. § 1045.

[37] 10 C.F.R. § 1045.4(a).

[38] 10 C.F.R. § 1045.6(b)(2)(iv).

E. Records Retention and Destruction

Because of the large estimated volume of classified materials, the Committee also reviewed the requirements of the Federal Records Act (FRA).[39] The FRA governs the creation and treatment of all federal agency records. It includes the Records Disposal Act (RDA), which governs the disposal and destruction of federal records.[40]

The FRA directs the Administrator of General Services and the Archivist of the United States, in coordination with agencies, to issue guidelines implementing the FRA's objectives for record management, which include "[a]ccurate and complete documentation of the policies and transactions of the Federal Government" and "[j]udicious preservation and disposal of records."[41] The Administrator and Archivist have general authority to examine agency records, except when records have restricted status "by law or for reasons of national security or the public interest." In such cases, inspection is subject to approval by the agency head or the President.[42]

Each agency head has a responsibility to "make and preserve records containing adequate and proper documentation of the organization, functions, policies, decisions, procedures, and essential transactions of the agency..."[43] Thus the FRA imposes an affirmative duty on agencies to create records.[44]

In addition, the agency head must set up an agency-wide program for effective records management and has the duty "to establish safeguards against the removal or loss of records he determines to be necessary and

---

[39] 44 U.S.C. §§ 2901 et seq.

[40] 44 U.S.C. § 3301.

[41] 44 U.S.C. §§ 2902(1)&(5), 2904.

[42] 44 U.S.C. § 2906(a)(2).

[43] 44 U.S.C. § 3101.

[44] Kissinger v. Reporters Committee for Freedom of the Press, 445 U.S. 136, 152 (1980).

required by regulations of the Archivist."[45] The agency head must notify the Archivist and initiate an administrative action through the Attorney General upon learning of "any actual, impending or threatened unlawful removal, defacing, alteration, or destruction of records" in agency custody.[46]

As part of the FRA, the RDA sets up exclusive procedures defining how and when agency records may be destroyed.[47] "If a document qualifies as a record, the FRA prohibits an agency from discarding it by fiat... [T]he FRA requires the agency to procure the approval of the Archivist before disposing of any record."[48] It directs the Archivist to promulgate regulations instructing agencies how to prepare schedules and lists of records for disposal. If, after examination, the Archivist determines that the records do not have "sufficient administrative, legal, research, or other value to warrant their continued preservation by the Government," the Archivist may, after publication of notice in the *Federal Register* and an opportunity for public comment, authorize disposal of the records by the agency.[49]

Nothing in the definition or the Act provides for special treatment of records containing RD or NSI. Thus, handling and destruction of such material is governed exclusively by the same language as that for nonclassified information. Indeed, the regulations implementing Exec. Order No. 12,356 (which presumably will also guide the implementation of Exec. Order No. 12,958) state that "[c]lassified information no longer needed in current working files or for reference or record purposes shall be

---

[45] 44 U.S.C. §§ 3102, 3105.

[46] 44 U.S.C. § 3106.

[47] 44 U.S.C. § 3314.

[48] <u>Armstrong v. Executive Office of the President</u>, 1 F.3d 1274, 1278-79 (D.C. Cir. 1993) (internal citations omitted).

[49] 44 U.S.C. § 3303a(a). The RDA has a provision for destroying records in an emergency (44 U.S.C. § 3310).

processed for appropriate disposition in accordance with [the FRA]."[50] If the Archivist approves the destruction of classified information, the agency "shall" destroy the records in accordance with procedures prescribed by the agency head. DOE has issued regulations for destruction of records that contain NSI and/or RD.[51]

****

In sum, the legal scheme imposes multiple, overlapping, and sometimes contradictory obligations governing the classification, retention, and release of documents. It is within this legal tangle that policy relating to classified information is currently being developed. The Committee hopes that both the government-wide reviews of the classification system and DOE's own efforts to enhance public access to information will result in significant rationalization and simplication of the legal structure.

---

[50] 32 C.F.R. § 2001.48.

[51] 10 C.F.R. § 1016.37.

CHAPTER 3

BASIC PRINCIPLES AND PRIORITIES FOR CHANGE

In reviewing current Department of Energy (DOE) policy and practice, the Committee recommends that DOE's approach to reevaluation of the classification system be guided by certain basic principles and that it give priority to certain legislative and regulatory changes.

A. Basic Principles

1. Minimizing the areas that are classified

Classification is clearly necessary when uncontrolled release of sensitive information could threaten national security. In a democracy, however, secrecy must be viewed as a sometimes necessary evil, to be used sparingly and only with strong justification. DOE's earlier study of classification policy, based on an extensive series of interviews throughout the DOE complex, concluded that "there is an almost universal belief that there is too much material to protect since some of it is now unnecessarily classified, or too highly classified. As a result, overclassification interferes with the protection of truly sensitive information."[1] In creating the new Commission on Protecting and Reducing Government Secrecy, Congress reached a similar conclusion:

> The burden of managing more than 6 million newly classified documents every year has led to tremendous administrative expense, reduced communication within the government and within the scientific community, reduced communication between the government and the people of the United States, and the

---

[1] Meridian Corporation, 1992, p. 86.

selective and unauthorized public disclosure of classified information.[2]

Congress further concluded that "if a smaller amount of truly sensitive information were classified, the information could be held more securely" and therefore directed the Commission "to make recommendations to reduce the volume of information classified and thereby strengthen the protection of legitimately classified information."[3]

The Committee agrees with the philosophy urged by Congress. We recommend that **the goal of DOE's current internal review should be to minimize to the maximum extent possible the subject matter areas that are classified**. DOE should seek to construct "high fences around narrow areas" -- that is, to maintain very stringent security around sharply defined and narrowly circumscribed areas, but to reduce or eliminate classification around areas of less sensitivity.

2. Shifting the burden of proof

As mentioned in the previous chapter, the Atomic Energy Act (AEA) creates a presumption that restricted data (RD) are classified merely by virtue of their existence without anyone having to make a classification decision -- they are "born classified." Without some affirmative act, all such information remains classified indefinitely. Many believe that since the enactment of the AEA in 1954, DOE and its predecessor agencies have used this authority to maintain secrecy about many of their activities, with little or no balancing of the harm that might result from disclosure with the detrimental costs -- to health, safety, scientific development, and democratic decision making -- that secrecy might create.

The AEA explicitly addresses declassification. It gives DOE responsibility to review "from time to time" whether RD "can be published

---

[2] Protection and Reduction of Government Secrecy Act, Pub. L. No. 103-236, § 902(3), 108 Stat. 525, 526 (1994).

[3] Protection and Reduction of Government Secrecy Act, supra, §§ 902(5), and 903(1).

BASIC PRINCIPLES AND PRIORITIES FOR CHANGE 41

without undue risk to the common defense and security."[4] The principle of disclosure within the AEA was largely confined to narrow areas during the Cold War; "undue risk" was assessed with a mind-set of absolute risk avoidance, where even the remotest threat had to be countered to the maximum extent. The requirement for declassification and disclosure now needs a broader interpretation.

The Committee recommends that in its comprehensive review of the basis for classification of information **DOE should shift the burden of proof from the proponents of declassification to the proponents of continued classification.** The Department should be guided by the presumption that information should not be classified unless there is an identifiable reason why release of the information could damage national security or a reason for concluding that the costs of release outweigh the benefits. Such a step would effectively reverse the burden of proof associated with RD from presumptively classified to presumptively unclassified. The burden would then be on the classifier to justify the classification of the information, rather than on those proposing declassification.[5]

3. Balancing costs and risks

Information should be classified only if the damage to the national security clearly outweighs both the public interest in disclosure of the information and the costs of attempting to prevent such disclosure. The Joint Security Commission noted,

> Security is a balance between opposing equities. The imperative to protect cannot automatically be allowed to

---

[4] 42 U.S.C. § 2162(a). Indeed, the Act encourages "dissemination of scientific and technical information relating to atomic energy . . . so as to provide that free interchange of ideas and criticism which is essential to scientific and industrial progress and public understanding and to enlarge the fund of technical information" (42 U.S.C. § 2161(b)).

[5] The U.S. Congress Office of Technology Assessment report makes a similar recommendation (Office of Technology Assessment, 1993, p. 162).

outweigh mission requirements or the public's fundamental right-to-know and it must never obscure the understanding that an informed public is the foundation of a democratic government.[6]

The Commission recommended a change in philosophy from "risk avoidance" to "risk management."[7] The Committee endorses these principles and urges the application of these concepts to information classified by DOE.

In deciding whether a given subject area should be, or should remain, classified, the Department should require that the benefits of classification clearly outweigh the costs.[8] Public acceptance of the approach would be enhanced if the balancing test were based on an agreed-upon set of criteria. Accordingly, the Committee believes that the criteria to be used in the declassification review should be developed with adequate opportunities for public input.

DOE currently has a set of criteria for use in reviewing information for possible declassification (see Table 2). The criteria are described as "typical of those which must be evaluated in determining whether publication would present 'undue risk' to the common defense and security." These criteria provide a good start, but the Committee believes that they **should be expanded to include explicitly the benefits of openness in enabling an informed public debate on public issues and, more generally, in enhancing the public's right to know what its government is doing.**

The Committee also believes that the public availability of information, which is included in DOE's current list of criteria, warrants careful consideration in declassification decisions. Some argue that there is no longer any point in maintaining the classification status of even

---

[6] Joint Security Commission, 1994, p. 4.

[7] Joint Security Commission, 1994, p. vi.

[8] The Joint Security Commission also observed that "[i]t is important to consider the political, economic, and opportunity costs of classifying information, as well as the costs of failing to classify information" (Joint Security Commission, 1994, p. 10).

Table 2

Department of Energy Declassification Criteria

1. The benefit to be realized by the U.S. program from the declassification action, including any significant technology commercialization potential.

2. The extent to which the information would assist in the development of a nuclear weapon capability in nonnuclear weapon states or in improvements to the weapons in a nuclear weapon state.

3. The cost in terms of time and money of acquiring the information.

4. The extent to which the information would assist in the production of special nuclear material.

5. The published state of the art for the information in the U.S. and other countries.

6. The cost to the U.S. program of the continued classification of the information.

7. Any detrimental (or beneficial) effect release of the information might have on U.S. foreign relations, arms control negotiations, or treaty obligations.

8. Any other national security impact or significance (e.g., the extent to which the information would assist an adversary nation assess or counter U.S. capabilities and limitations).

9. Any impact on the credibility of the DOE classification program of the continued classification of the information.

After U.S. Department of Energy Office of Declassification, 1994a, p. 7.

weapons-design information that has already been made available to the public. Others argue that information that is unclassified and might be "available" to sophisticated and experienced experts who already understand its significance may not be effectively available to less sophisticated potential proliferators, and that DOE should not publicize the existence or confirm the validity of any such information. In other words, even if material has appeared in the news media, declassification is not necessarily indicated since this would serve to validate such information. In the Committee's view, **public availability of information should be an important consideration, although this factor should not be the prevailing or overriding criterion for declassification of that information.**

4. Enhancing openness and public access

Much can be done to avoid both the reality and the perception that classification authority is being used to withhold information that the public has a legitimate right to know. Classification policies that are understandable to external parties are more likely to be supported, or at least accepted, than those that are obscure or hidden. Furthermore, such understanding and support are more likely if outside parties are allowed to participate in an informed way in the development of those policies than if they are imposed without consultation. The Openness Initiative and the various reviews of declassification policy currently under way offer a number of opportunities for practical steps to demonstrate the Department's commitment to increased public access to information. **DOE's goal should be "open policies openly arrived at." To the maximum extent possible, the debates about new information-control policies should be open to the public, with ample and credible opportunities for public inputs.**

DOE needs to provide for formal external input in the review process. To ensure that the costs and benefits of disclosure are given appropriate weight, **DOE should seek formal input to the comprehensive review of classification policy by those who are affected by the decisions.** This would be a direct demonstration of the new DOE spirit of openness and would increase the likelihood that the revised policy will be understood and accepted by those outside the security community. **DOE**

should establish an Information Policy Advisory Board, appointed by the Secretary and composed of experienced outside experts broadly representative of the major stakeholders in DOE's classification policy. The board would initially provide systematic external input to the current fundamental review. Later it could serve a variety of functions, such as making recommendations on priorities for document declassification efforts.

The Advisory Board should report to the DOE official charged as the senior operating officer; at present this is the Under Secretary. Board members should have appropriate clearances to enable them to participate fully in discussions and reviews.

Such a Board could provide a means for informed and direct discussions between DOE and knowledgeable outsiders on a more continuing and intensive basis than is possible with single public meetings. It would also facilitate direct discussion among the affected communities, which could provide useful insights about balancing different points of view in establishing the new policy.[9] It could perform a variety of other functions as well, which are discussed in later chapters; for example, the Board could render nonbinding but public recommendations to DOE concerning the justification for classifying specific categories of information or for deciding not to declassify information in response to a request for declassification.

Because it may take some time to establish the Board, the Department should seek in parallel suggestions from outside parties about the criteria for the declassification effort and areas of information that are candidates for declassification. This would extend to outside parties the current biennial call to DOE elements for proposals for declassification. One possible mechanism for this input would be publication of notices of proposed rulemaking, draft rules or guidance, and similar policy documents in the *Federal Register* for public comment. The Office of Declassification should be given responsibility for ensuring that these suggestions are given

---

[9] The Joint Security Commission called for "a security advisory board composed of distinguished Americans who would provide a non-government and public interest perspective to security policy. The board would act as a barometer for the committee to ensure that security policy and implementation is consistent with the overall goals of the government, such as openness, cost effectiveness, and fairness" (Joint Security Commission, 1994, p. 129).

a fair evaluation in the review process, since uncleared outsiders may be at a serious disadvantage in supporting their comments under circumstances in which they do not have access to all the relevant information.

When the joint review is completed, DOE should indicate publicly which areas of information it believes no longer require protection as RD. Even if the Department of Defense (DOD) or other agencies object to declassification on other grounds, public understanding would be enriched by knowing whether the basis for continued classification has to do with the military sensitivity of the information (that is, its military value to a potential adversary of the U.S. or its allies and friends) or with its possible value to those seeking to construct a nuclear weapon.

DOE should promptly release a final version of its report entitled "Public Guidelines to Department of Energy Classification of Information." The Committee commends DOE's release of a draft of this report for public comment on June 27, 1994. The document should serve to relieve the concern that the public has been insufficiently informed about the boundaries between unclassified and classified information, and that in fact much of the knowledge of these boundaries is in itself classified. The Committee is gratified that the Secretary accepted the suggestion made at its February 1994 work session for preparation of this guide, although it understands that the final version has not yet been issued.

Release of the draft document -- and its finalization and amplification -- should greatly facilitate informed public involvement in the classification policy review. This document should contain, for each area of information that remains classified, a rationale for that decision, including a discussion as to how the cost and benefit criteria were applied. This need not entail a great deal of additional work, since such a rationale should be prepared in any case as part of the review process. It should also serve to reduce the volume of information requested by the public, because it enables the information requester to determine what subject areas are classified. To further these objectives, the Department should make the draft and final versions available through OPENNET, the computer-based information network being set up by DOE to disseminate information about declassified documents.

B. Priorities for Legislative and Regulatory Changes

DOE should move expeditiously with its effort to review and reform its classification system. The Secretary of Energy has unique authority with respect to the control of nuclear information, and **should proceed to use that authority to reform the Department's policies and practices in this area.**

While DOE should aim toward compatibility with the evolving government-wide approach to classification, it should not slow its efforts to match the pace of what is necessarily a more cumbersome interagency review and negotiation process. **DOE should also continue to take the lead in seeking declassification of information about nuclear weapons that it believes can be released without undue risk, but is subject to some degree of control by other agencies.**

1. Amending the Atomic Energy Act (AEA)

A thoughtful study has been prepared for DOE of possible legislative changes to the AEA,[10] including a set of proposed amendments. While DOE notes that "it is the Department's classification and information control policies under the Act that influence the public's perception of the Department," it adds that "amendments would make the Department's commitment to openness in dealing with the public's interests a matter of law."[11]

The Committee supports a careful review of the AEA in light of the changing security needs of the nation. The Committee did not attempt to review and critique all the recommendations from the Meridian Corporation study for DOE on possible amendments to the AEA, and thus it does not offer firm conclusions about them.

The Committee endorses one recommended amendment. DOE shares with DOD the authority to remove some information primarily related to military use of atomic weapons from the category of RD if the agencies determine it can be adequately protected as national security

---

[10] Meridian Corporation, 1992.

[11] U.S. Department of Energy, 1994, p. 8.

information. This category of information is called formerly restricted data (FRD). Unlike national security information (NSI), however, FRD cannot be transferred to any other country except as part of an agreement authorized as part of the AEA. Because this constraint appears needlessly confining, **DOE should seek legislative authority to simply transclassify to NSI any RD that no longer warrant special protection as nuclear-related information but still may be sensitive for other military or diplomatic reasons, thus permitting elimination of the entire category of FRD.**[12] Elimination of the FRD category and transclassification to NSI would eliminate some international complications caused by the requirements for controls over FRD.[13] It would also subject such information to Exec. Order No. 12,958 requirements for systematic and mandatory declassification reviews and possible automatic declassification after a specified period or event.

Other proposed amendments to the AEA might be considered as an integral part of the broad two-year review of classification policy being undertaken by the Commission on Protection and Reduction of Government Secrecy created by Congress in 1994. For example, that Commission would be a proper venue for consideration of the fundamental question of whether there is any continuing justification for two separate and parallel classification systems, one for RD controlled pursuant to the AEA and one for NSI controlled pursuant to executive order. The systems are similar and the redundancy means additional cost. The separate systems contribute to inefficiency in interagency activities because documents classified under the aegis of DOE and those classified by other governmental departments are frequently difficult to exchange. Similarly, the clearance processes for personnel duplicate one another. Establishing a common classification system that applies across all agencies could facilitate more efficient and effective government.

---

[12] These are data that DOE and DOD jointly determine relate primarily to military applications of nuclear weapons and that could be protected as defense information. The Classification Policy Study (Meridian Corporation, 1992) recommends amendment to the AEA to eliminate the category of FRD and to transfer military utilization information to NSI.

[13] Meridian Corporation, 1992, p. 55.

We recognize that there are possible costs associated with a unified system. DOE's largely autonomous authority over RD enables the Secretary to develop policy without the burden of a cumbersome and slow interagency process. Preservation of the existing system thus may have short-term benefits in a time of rapidly evolving policy, but almost certainly there would be long-term gains from a uniform set of government policies and procedures.

DOE should not wait for amendments to the AEA to implement desired openness policies that are allowed by the Act. The Department points to the major declassification actions already taken by the Secretary as evidence that significant steps can be taken without any legislation.[14] The Committee notes that some of the specific amendments proposed by DOE can, and should, be incorporated directly in DOE policy now.

One important policy change would be **establishing a systematic declassification review of existing documents containing RD, based on priorities reflecting public needs and interests, and on available resources.**[15] Exec. Order No. 12,958 requires that classified documents containing NSI be marked with a date or event for automatic declassification of the document or the category of exemption from declassification that applies to the document.[16] Since much nuclear weapons information does not become less sensitive with the passage of time, there are no provisions for automatic declassification of documents containing RD and FRD. The AEA already requires such reviews of categories of information "from time to time."[17] DOE's study of amendments to the AEA has proposed an amendment requiring such systematic reviews. DOE could adopt such a policy on its own, pending any amendment. Such reviews should be based on priorities reflecting public needs and interests and on available resources.[18]

---

[14] Meridian Corporation, 1992, p. 8.

[15] Meridian Corporation, 1992, p. 14.

[16] Exec. Order No. 12,958, § 1.7(a)(4); exemptions are defined in § 1.6(d).

[17] 42 U.S.C. § 2162.

[18] U.S. Department of Energy, 1994, p. 14.

A second important change relates to prohibition of **abuses of classification or the control mechanisms established for RD and FRD.** Exec. Order No. 12,958 already forbids use of classification as NSI "in order to conceal violations of law, inefficiency, or administrative error; prevent embarrassment to a person, organization, or agency; restrain competition; or prevent or delay the release of information that does not require protection in the interest of national security."[19] Essentially identical prohibitions are also contained in DOE's regulation concerning unclassified controlled nuclear information (UCNI).[20] However, there are no such prohibitions with respect to RD and FRD in DOE's basic order governing classification.[21] Such prohibitions should extend to RD and FRD.[22] Further, it should be made clear that <u>delay</u> of declassification of RD or FRD for any of the prohibited reasons is also an abuse.

The Committee perceives no reason why the Secretary cannot make these changes without seeking legislative action and recommends that they be made.

2. Using the regulatory process

The Committee recommends that, **where possible, DOE should develop and adopt new rules and procedures as regulations issued under the authority of the AEA and pursuant to the Administrative Procedures Act (APA).** The APA has provisions for notice and public participation in many agency rulemaking situations. DOE already has

---

[19] Exec. Order No. 12,958, § 1.8(a).

[20] 10 C.F.R. § 1017.5.

[21] The Policy and Objectives section of Order DOE 5650.2 (Chapter III) (U.S. Department of Energy Office of Classification, 1991) does contain the prohibitions on misuse of NSI classification, but there are no equivalent requirements concerning RD and FRD.

[22] The Department's study of amendments to the AEA proposes an amendment to prohibit the abuse of classification or UCNI controls (U.S. Department of Energy, 1994, p. 10-11).

promulgated regulations under the AEA dealing with NSI and UCNI,[23] but has none dealing directly with RD.

The embodiment of new classification policies in rules promulgated under the APA would have several significant advantages. First, such rules would largely replace the current system of DOE orders. Rules would provide more stability than DOE orders, since revisions require a new rulemaking and public explanation of the reasons for changes. While less permanent than legislation, such rules are not subject to the vagaries and delay associated with the legislative process; within the constraints of current law, DOE can proceed with rulemaking on its own initiative and schedule.

Second, use of a rulemaking process is consistent with the goal of "open policies openly arrived at." A rulemaking provides a well-understood and accepted mechanism for formal public input, which should increase understanding, and potentially acceptance, of DOE classification control policies. It also increases accountability, since decisions must be explained and are subject to judicial review.

Finally, rulemakings now could facilitate legislative revisions later. The rulemaking process would identify areas where amendments to AEA are required, rather than simply helpful clarifications. It could also develop a record of stakeholder views that could facilitate legislative changes when the time comes and language that could be incorporated into clarifying amendments.

Specifically, DOE should promulgate a new regulation concerning classification and declassification of RD. A rulemaking to promulgate a regulation for RD could proceed in parallel with the fundamental review of classification policy, since the latter is focused more on what information should be classified than on the broader structure and procedures of the classification system. Such a rulemaking would also provide a good vehicle for public discussion of the recommendations of this report and those of the parallel studies now under way.

---

[23] The UCNI regulation is found at 10 C.F.R. § 1017; the NSI regulation is found at 10 C.F.R. § 1045.

CHAPTER 4

ISSUES IN CLASSIFICATION POLICY

Classification policy and the classification guides apply to categories and types of <u>information</u>. Once the information that is to be classified has been generated and defined, the task of classifiers and declassifiers is to determine whether a particular <u>document</u> contains classified information. Building on the general principles discussed in the previous chapter, this chapter addresses the question of what information should be classified. The difficult issues of identifying specific documents containing the information of interest and having them declassified and disseminated are the subject of Chapter 5.

There are public pressures for openness in four main areas: (1) the effects of Department of Energy (DOE) activities on health, safety, and the environment; (2) the exploitation of classified technologies with potential commercial applications;[1] (3) the historical actions of DOE and its predecessor agencies (the Atomic Energy Commission and the Energy Research and Development Administration); and (4) nuclear weapons policy, dismantlement of surplus weapons, and management of the resulting materials. Significant advances have been made in declassifying <u>information</u> in the first two areas; some progress has been made in the third and fourth. All information related to health, safety, and the environment has been declassified in principle.[2] Similarly, declassification of information with significant potential for commercial application has repeatedly occurred since the early days of the Atomic Energy Commission. Examples include declassification of information relevant to development of civilian reactors, information on reprocessing, and recently, most information relating to inertial confinement fusion. Significant issues remain to be addressed, however, especially regarding nuclear weapons information.

---

[1] U.S. Department of Energy Office of Declassification, 1994b.

[2] U.S. Department of Energy Office of Declassification, 1994a, p. 40.

A. The Case of Nuclear Weapons Information

1. General policy issues

A well-informed national policy debate concerning the military application of nuclear weapons, nuclear nonproliferation, dismantlement of surplus weapons, and disposition of the resulting materials serves the public interest. We believe that public participation in such policy deliberations is necessary to obtain the broad public acceptance that will be needed to allow revised policies to be implemented effectively.[3] But effective participation will not be possible unless the necessary information is declassified and made available in a timely manner. The Committee agrees with the view of President Reagan's Blue Ribbon Task Group on Nuclear Weapons Program Management:

> One of the national security responsibilities of DOE leadership is to make available sufficient information to allow informed public debate on nuclear weapon issues. The Task Group urges that DOE review its classification procedures to ensure that criteria are based upon current requirements rather than historical precedent.[4]

DOE has only taken initial steps toward the declassification of information that will help to inform the public debate about nuclear weapons policy. Both the Secretary of DOE and the Director of the Office of Declassification have acknowledged the need to declassify more such information, insofar as possible, and on December 7, 1993, and June 27,

---

[3] "[P]olicies developed entirely behind closed doors are unlikely to achieve public acceptance... For effective policy development, information access will have to be enhanced and participants in the debate will have to come from more sectors of government and society than in the past" (Office of Technology Assessment, 1993, p. 122).

[4] President's Blue Ribbon Task Group on Nuclear Weapons Program Management (1985, p. 13). This conclusion was also endorsed by the National Academy of Sciences Committee on International Security and Arms Control (1994, p. 91).

1994, the Secretary announced the declassification of certain additional categories of information. Although progress has been significant, the fact remains that some items of information having important bearing on weapons policy debates remain classified. In view of the changed world situation, the Committee concludes that consideration of further declassification actions is warranted.

Some of the areas in which there has been limited disclosure and that should be reexamined with a view to further declassification include

a) Yields of nuclear tests carried out since 1962. Knowledge of such yields would be extremely unlikely to provide information to potential proliferators about weapons design, but release of the data would serve U.S. national interests by making possible the calibration of seismic detection networks, thereby paving the way to more extensive foreign collaboration in nuclear test detection.

b) Information on weapons stockpiles and stockpiles of special nuclear material. A potential proliferator would in no way be influenced by the precise knowledge of these figures, which are enormously in excess of any aspirations of a proliferator.

c) Actions of DOE and its predecessors that are of significant historical importance. More aggressive declassification would greatly aid historical reconstruction of crucial actions regarding nuclear weapons policy, research, development, testing, and production and would be of invaluable assistance to scholars and policy analysts seeking to learn the lessons of the Cold War.

While the declassification of documents containing recently classified information, at least on a large scale, is a substantial and time-consuming undertaking, it should be possible to release information of major interest in summary form. For example, a request for information about test yields is in fact a request for specific information, not for documents. Releasing such information would show that the Department is committed to ending the practice of blanket secrecy on such matters, substituting instead a practice of releasing information that enriches public debate, while continuing to withhold information that contains legitimate national security secrets. If serious questions about the accuracy of the information were to arise, the Department could ask the Information Policy Advisory Board to review the classified source documents to verify the accuracy of the information.

Such an approach might satisfy some of the public demand for information without the massive effort that declassification of the inventory of files will require. This approach, however, will not obviate the practical and legal obligation to produce documents eventually. Historians, policy analysts, and others understandably will demand the opportunity to review actual files. Some requesters, even if given information, will ask to see the actual documents to be sure that the information supplied is accurate and complete. Thus, the approach of providing information cannot be a substitute for the effort of declassifying documents. Indeed, the Freedom of Information Act (FOIA) imposes obligations to produce documents, not just information. Nonetheless, it offers a way to begin the process and may reduce the number and scope of FOIA demands.

2. Transparency vis-à-vis the Russians

At their summit meeting in January 1994, Presidents Bill Clinton and Boris Yeltsin agreed on the goal of ensuring the "transparency and irreversibility" of the nuclear arms reduction process.[5] In addition, the United States is currently engaged in negotiations with Russia to persuade it to strengthen controls over the management of special nuclear material, as well as over spent nuclear fuel from the commercial fuel cycle. This urgent problem was highlighted by the recent, highly publicized smuggling of special nuclear material from Russia. A separate National Academy of

---

[5] White House, "Joint Statement by the President of the Russian Federation and the President of the United States of America on Non-Proliferation of Weapons of Mass Destruction and the Means of Their Delivery," January 14, 1994, p. 2. "Transparency" refers to a code of conduct by which information about one's actions is fully accessible to others.

Sciences report addressed these problems in considerable detail.[6] One major recommendation of that report is to greatly increase the public availability of information both in Russia and the United States concerning the content, location, and management of stockpiles of strategic and other nuclear materials.

Congressional actions clearly support release of nuclear weapons-related information on a reciprocal basis with Russia. The 1992 Senate resolution approving ratification of START I included the "Biden Condition," which required the President to seek arrangements to monitor stockpiles of weapons and fissile materials in the United States and the former Soviet Union using "reciprocal inspections, data exchanges, and other cooperative measures."[7] The Defense Authorization Act for FY1993 permitted declassification of stockpile information in the context of an agreement between the United States and Russia for release of such data.[8] As a further step in this direction, the Defense Authorization Act for FY1995 allows a one-year period (until December 31, 1995) in which the United States might reach an agreement for exchange of classified nuclear weapons information with Russia without requiring the full congressional review process provided by the Atomic Energy Act (AEA) for such international agreements.[9]

The Secretary's Openness Initiative has gone a considerable way toward the goal of transparency, and Russia has at least initially accepted some proposals to provide greater openness. For example, in March 1994 DOE Secretary O'Leary and Russian Minister of Atomic Energy Viktor Mikhailov reached agreement on unprecedented inspections of fissile

---

[6] National Academy of Sciences Committee on International Security and Arms Control, 1994.

[7] National Academy of Science Committee on International Security and Arms Control, 1994, p. 93.

[8] National Academy of Science Committee on International Security and Arms Control, 1994, p. 91.

[9] National Defense Authorization Act for FY1995, Pub. L. No. 103-337, § 3155, 108 Stat. 2663, 3091-92 (1994).

materials from dismantled nuclear weapons.[10] During their discussions in June 1994, Vice President Al Gore and Russian Prime Minister Viktor Chernomyrdin agreed to establish a joint working group to explore the confidential, reciprocal exchange of data on stockpiles of nuclear weapons and fissile materials. This was followed in September by a Clinton-Yeltsin commitment to exchange stockpile data by the end of 1994.[11]

As of early 1995, that exchange has not taken place. Russian secrecy, particularly in the nuclear energy field, is deeply ingrained and is yielding slowly and reluctantly to pressures for change. On the U.S. side, the delay lies in part in the need to achieve interagency agreement, but also on the strictures of the classification system. Much of the stockpile data that would be exchanged with Russia is restricted data (RD) or formerly restricted data (FRD). The U.S. has tabled a draft Agreement for Cooperation, but until that is signed by both countries, the data exchange cannot take place.

Overcoming the reluctance in both countries to reveal information long held secret requires sustained, high-level attention. **DOE should continue to pursue reciprocal exchanges of information with Russia. Specifically, the Department should explore arrangements in which each party to the exchange retains the right to allow or prevent the public release of the information that it is providing to the other party, so that disagreements about whether information should be publicly released do not obstruct mutually beneficial exchanges.**

The Committee concurs with the Academy's report on management and disposition on excess plutonium weapons about the urgency of measures to improve control and accounting of Russian fissionable materials and to improve transparency about the nuclear stockpiles of both the U.S. and Russia. **However, the focus on this important objective should not be allowed to delay the release of declassifiable information about American nuclear weapons that is needed to enable informed debate about policies appropriate to the new era.** The pursuit of a

---

[10] Negotiations over the sites and rules for inspections were still under way in early 1995, but when they begin, the inspections will offer unprecedented access for each side to the other's nuclear facilities.

[11] White House, "Joint Statement on Strategic Stability and Nuclear Security by the Presidents of the United States and Russia," September 28, 1994, p. 3.

confidential exchange of RD or FRD with Russia should not serve as a justification for delay in declassifying and releasing U.S. data.

B. The Special Case of UCNI

In 1981 Congress gave the Secretary of Energy authority, under section 148 of the AEA, to prohibit the unauthorized dissemination of a category of information designated unclassified controlled nuclear information or UCNI. The AEA defines UCNI to include information pertaining to

a) the design of production facilities or utilization facilities;
b) security measures concerning such facilities or nuclear material that is contained in such facilities or is in transit; or
c) the design, manufacture, or utilization of any atomic weapon or component if that information has previously been declassified or removed from the restricted data category.[12]

DOE can designate information in these areas as UCNI if it determines that "the unauthorized dissemination of such information could reasonably be expected to have a significant adverse effect on the health and safety of the public or the common defense and security by significantly increasing the likelihood of (A) illegal production of nuclear weapons or (B) theft, diversion, or sabotage of nuclear materials, equipment, or facilities."[13] Although UCNI is unclassified, it can be shared only with U.S. citizens in specified categories having a need to know, and failure to abide by the restrictions is subject to stiff sanctions, including a substantial civil penalty.

---

[12] 42 U.S.C. § 2168(a)(1)(1988).

[13] 42 U.S.C. § 2168(a)(2).

UCNI is controversial.[14] Critics see the category as vague, with inclusion criteria that are difficult to distinguish from the criteria for classification.[15] They charge it is being used too widely despite the strictures in the AEA to minimize its use. UCNI may also be subject to abuse; for example, it offers an easy way to avoid release of information requested under the FOIA.[16] Critics argue that it creates barriers to information flow without clearly providing much protection against proliferation, adds cost and bureaucracy, and, in general, diverts resources from protection of more important information.

On the other hand, supporters argue that it is needed to protect sensitive information that could not otherwise be classified for legal or operational reasons. They also claim that it allows wider dissemination of information of relatively low sensitivity than would be possible if the only alternatives were classification or complete declassification.[17]

The Committee has not received persuasive justification for the retention of UCNI. The concept of information control beyond classification runs counter to the objective of focusing resources on protecting truly sensitive information. Moreover, the application of UCNI over the years seems to have drifted some distance from the language of the AEA and DOE's implementing regulation.[18] As discussed below, the circumstances and context within which UCNI is applied have changed

---

[14] Other forms of DOE control over unclassified information are also controversial. For discussion, see Adler (1993); Meridian Corporation (1992), p. 62-68; and Shinn (1990).

[15] The definition of "confidential" information in Exec. Order No. 12,958 is "information, the unauthorized disclosure of which reasonably could be expected to cause damage to the national security." The use of the word "significant" in the UCNI provision noted in the above text, in contrast to the lack of any such qualifier in the definition of "confidential" information, could be read as suggesting that anything passing the UCNI test is at least as sensitive as confidential information.

[16] Meridian Corporation, 1992, p. 65; Adler, 1994.

[17] U.S. Department of Energy, 1994, p. 7.

[18] See, for example, U.S. Department of Energy, 1994, p. 21-22.

sharply since the amendment was passed and DOE regulations developed. **The Committee recommends a thorough reexamination of the need for UCNI. The Congressional Commission on Protecting and Reducing Government Secrecy or other appropriate body should reassess UCNI in the context of a broader review of controls on unclassified information.**

1. The original purpose of UCNI

In 1981 the principal reason for creating UCNI was to control information about sensitive nuclear facilities that might be of use to terrorists.[19] This information includes layouts for DOE buildings showing where radioactive materials are stored and where critical operations are conducted and emergency response plans at such facilities. Many uncleared people, such as police or firemen, might need such information to perform their jobs, but DOE does not consider it feasible to classify the large volume of documents involved and to provide security clearances to all people whose jobs require access to them.[20] Placing such information in the UCNI category enables DOE to make the information available without the cost, administrative burden, and delay of security clearances and classification procedures. At the same time, the availability of the information can be limited because of the penalties that can be imposed for unauthorized dissemination and because the statutory language provides a clear exemption from FOIA requests.[21]

---

[19] Statement of Considerations for 10 C.F.R. § 1017, Identification and Protection of Unclassified Controlled Nuclear Information, 50 Fed. Reg. 15,818 (1985).

[20] "Unclassified Controlled Nuclear Information," a two-page paper dated February 3, 1994, and subtitled "Unclassified but Sensitive Information" (no author identified), (provided to the Committee by the DOE Office of Declassification) is the only document the Committee has seen that states that UCNI is used for information that cannot be classified for operational or cost reasons.

[21] The UCNI provision is modelled on an earlier amendment to the AEA giving the Nuclear Regulatory Commission (NRC) the authority to control, and deny

(continued...)

According to the statutory provision described above, UCNI controls can be applied to information about the design of production and utilization facilities (i.e., nuclear reactors). These controls do not clearly apply to facilities for manufacture, assembly, or storage of nuclear weapons or their components -- the facilities of greatest current concern. However, the Department has interpreted section 148 broadly to include such facilities, on the grounds that Congress intended section 148 to cover information related to nuclear weapons. DOE is considering a recommendation that the AEA be amended to clarify that all nuclear weapons facilities are covered.[22]

The more fundamental question is whether UCNI is needed at all to protect information related to facility security. None of the studies addressing UCNI that were provided to the Committee contained any quantitative analysis that supports the argument that it is impractical to control security-related information about facilities through classification as NSI.[23] The Committee recommends that **DOE evaluate the costs and feasibility of treating sensitive information related to facility security either as a special category of national security information (NSI) or, alternatively, as unclassified information not subject to special controls.**

(...continued)
FOIA requests for release of, information concerning applicants' and licensees' safeguards and security provisions. A major motivation for that earlier amendment was NRC's concern that without such statutory authority, it could not restrict release of such information. [Adler (1994), p. 87; 10 C.F.R. § 1017: Identification and Protection of Unclassified Controlled Nuclear Information 50 Fed. Reg. 15,818 (1985)].

[22] U.S. Department of Energy, 1994, p. 21-22.

[23] Such information would be classified as NSI rather than as RD (Meridian Corporation, 1992, p. 65). It should be noted that the claim that information that is currently UCNI could be classified as NSI appears to be inconsistent with DOE's order implementing UCNI, which says that "UCNI controls should not be used in place of classification if classification is appropriate" (U.S. Department of Energy Office of Security Affairs, 1992, Order DOE 5650.3A, § 8.g).

2. New uses of UCNI

More recently, the growing emphasis on proliferation and the control of information in proliferation-related areas such as plutonium processing, isotope separation technologies, and high explosive technology related to nuclear weapons has expanded the range of information considered UCNI.[24] Although the original focus of UCNI was protecting information relating to the physical security of sensitive facilities, its use has evolved to control a broad spectrum of information on dual-use technology -- that is, technology that can support both military and civilian applications.

DOE's study of classification policy raises questions about its authority to extend UCNI controls to proliferation-related information. The AEA focuses on "illegal production of nuclear weapons." This phrase does not clearly apply to production of weapons by sovereign nations or even by subnational groups outside the United States.[25] Moreover, the definition has important areas of ambiguity; the reference to "the design of production facilities or utilization facilities" could be interpreted as applying only to information about existing facilities that could be useful to a terrorist in planning an attack, not to more abstract information that could help a proliferator construct and operate such facilities.[26] **If UCNI**

---

[24] Meridian Corporation, 1992, p. 65.

[25] Recognizing this problem, a draft DOE study proposes amending the AEA to clarify the applicability of the UCNI provisions to discourage proliferation by removing the adjective "illegal" in reference to production of nuclear weapons (U.S. Department of Energy, 1994, p. 21-22). In the meantime, DOE guidelines interpret "illegal" as meaning contrary to domestic or international law. (U.S. Department of Energy Office of Classification, 1993, p. III-3).

[26] In 1988, years after DOE's UCNI regulation was promulgated, DOE issued an important interpretation identifying two separate categories of information about "design": "technological design information" that is independent of the technical details of specific existing facilities but that could assist a proliferator in constructing facilities for producing and preparing nuclear materials for weapons, and "security-related design information" dealing with details of designs and design-related operational measures that would aid a saboteur or

(continued...)

is to continue to be used as the basis for controlling proliferation-related technical information, the critical interpretations that provide the basis for that use should be included in an updated UCNI regulation.

While Exec. Order No. 12,958 provides for reclassification of NSI that has been declassified, the "traditional" interpretation of the AEA by DOE's Office of General Counsel has been that once <u>an area</u> of RD has been declassified, even new information in that area cannot be classified as RD, FRD, or NSI unless there was a caveat in the original declassification decision that contemplated such an action.[27] Thus, once an area is declassified, subsequently developed information in that area is not eligible for classification, even if it is highly sensitive. UCNI is understood, however, to allow the recapture of information that has been declassified. This opportunity is offered as a justification for preserving the UCNI category.

In our view, UCNI should not be used for classification of information that is highly sensitive, since the level of protection that can be afforded by UCNI controls is much lower than for information classified RD or NSI. In apparent recognition of this, DOE proposes an amendment to the AEA to allow classification of new information in areas that have been previously declassified.[28] Before seeking an amendment to the AEA, however, DOE should carefully reexamine the "traditional interpretation" that the AEA precludes such action. If a reinterpretation is possible, the proposed regulation dealing with RD discussed below should include a section providing for classification of new information in broad areas of information that have been declassified, based on a clear demonstration that uncontrolled release of this new information would cause an "undue risk" to national security. If a more favorable interpretation is not reached,

---

(...continued)
    thief in attacking an existing DOE facility (P. LaPlante, DOE Office of Declassification, personal communication, September 19, 1994).

[27] U.S. Department of Energy, 1994, p. 20.

[28] This is distinct from "reclassification," since the specific information in question was never directly declassified; only the broader subject area and the information it contained at the time was declassified (U.S. Department of Energy, 1994, p. 20).

ISSUES IN CLASSIFICATION POLICY 65

DOE should seek an amendment to the AEA to provide the authority to reclassify. In any case, DOE should not rely on UCNI to protect truly sensitive information.

Finally, DOE states that UCNI provided gradations in the control system that allow broader but not unlimited dissemination of information.[29] DOE argues that an "all-or-nothing" classification scheme denies the opportunity to make the most appropriate use of information and that the change would be at least a partial move in the direction of the Department's Openness Initiative. It asserts that UCNI is used to serve a valuable purpose by allowing information of a lower level of sensitivity to be shared with the private sector for purposes such as commercialization, while still affording a degree of protection against nuclear proliferation.[30]

UCNI seems poorly suited to this purpose. DOE's order implementing UCNI prohibits its use in place of classification controls under the AEA or executive order if classification would be appropriate and prohibits information from going directly from the category of RD to UCNI.[31] Under these restrictions, the only way for nuclear information that had been classified as RD to be controlled under UCNI is for the information to be declassified and then for circumstances to change such that the information becomes sufficiently sensitive to warrant reassertion of some degree of control. Consequently, the current UCNI provisions do not appear to be particularly well suited to the objective of providing a "middle ground" in the classification spectrum.

---

[29] U.S. Department of Energy, 1994, p. 15.

[30] "The Unclassified Controlled Nuclear Information system recognizes that there is a spectrum of sensitive information which requires a consistent spectrum of protection. Since Unclassified Controlled Nuclear Information is less sensitive than classified information, its protection and access requirements are also less stringent... If the only option to protect sensitive information were to classify it, much information of lesser sensitivity would be overprotected and denied to those who have a legitimate need to have it" (U.S. Department of Energy, 1994, p. 7).

[31] DOE Order 5650.3A Identification of Unclassified Controlled Nuclear Information, § 8(g) (U.S. Department of Energy Office of Security Affairs, 1992).

An amendment to the AEA, or perhaps to the DOE order, might remedy this limitation, but the need to add intermediate levels of information control should be examined carefully before they are adopted. The creation of new categories of control is contrary to the basic premise of an effective system -- tight controls over narrow areas.

If DOE concludes that information now encompassed by UCNI should continue to be protected under this scheme, it should prepare a clear and thorough background information document describing and explaining the rationale for the proposed uses of UCNI and a comparison of alternative approaches to achieving the same objectives. Currently available documentation is inadequate for that purpose. It is difficult, if not impossible, for an outsider, even with some diligence, to obtain a clear picture of the scope, application, and rationale for UCNI from available documents. The legally required quarterly reports on UCNI decisions provide little insight into the underlying rationale for the decisions. Furthermore, the two DOE classification policy studies provided to the Committee do not give an adequate discussion of options to allow the uninitiated to form an opinion.[32] DOE cannot expect to be granted the benefit of the doubt by skeptics when it is difficult to obtain a clear picture of what it is doing in this area and why it is doing it.

---

[32] These two studies (Meridian Corporation, 1992; and U.S. Department of Energy, 1994) are far too cursory and incomplete to allow an informed judgment about the need for UCNI; they report positions of various parties, but contain little or no analysis of the underlying facts and issues.

CHAPTER 5

DECLASSIFYING DOCUMENTS

A narrowing of the <u>information</u> subject to classification is not itself sufficient to achieve a policy of openness. The Department of Energy (DOE) must find a means for reviewing <u>documents</u> to determine whether they contain only unclassified information and, if so, for releasing them to an often skeptical public. This chapter reviews the process of handling classified documents, including declassification, and suggests a number of both general and specific improvements. Newly generated documents, which are subject to new procedures, policies, and guidelines, are discussed separately.

A. Dealing with Existing Documents

The changes in policy discussed in Chapters 3 and 4 should serve to minimize overclassification of newly generated documents and to establish a framework for declassification of old documents. But changes in policy must lead to processes for screening and releasing old documents as well.

The Committee encourages efforts by DOE to declassify and publicly release many of the documents that have accumulated over the past 50 years. This action will benefit many groups: scientists who can apply the basic data in their own work; engineers working on problems that may have been solved in connection with classified work; historians interested in details of the past confrontation between the two former superpowers and in U.S. nuclear science and technology programs; DOE site employees tasked with characterizing buried wastes and other contaminants; and members of the general public concerned about the effects of nuclear materials and processes on society, health, safety, and the environment.

1. Defining the problem

Even after information has been declassified, the Department still faces a significant task in making documents containing that information available. The constant challenge for the Department is to identify and then locate the documents of interest to the inquirer, determine whether they are commingled with classified information, and -- if not -- make them available.

The first aspect of the challenge is simply the huge quantity of classified documents. As noted earlier, according to DOE's current estimate, there are some 280 million classified pages. This estimate is larger by a factor of 10 than the estimate made in late 1993, and it would not be surprising if the estimate continues to increase. Moreover, no master index of historical documents (not even a simple title index) exists, making it difficult for researchers to know what information -- classified or not -- might exist and where it is located. The Department is currently conducting an inventory of <u>all</u> records, classified and unclassified. This will be at the <u>series</u>, not the individual <u>document</u>, level. (It will include title, dates, general information content, and the highest classification level of documents in series.)

Locating a relevant document is only the first step in its release. As discussed in earlier chapters, a classified document that contains information that has been declassified may contain other information that is still classified, or might be commingled with other still-classified documents in files, boxes, or storage areas. The Department must review the document in question to determine whether it contains any information that is still classified. Under the present system, this is a labor-intensive process that appears ill suited to any effort to declassify documents on a large scale.

Some 200 individuals in the Department currently have declassification authority. As part of the Secretary's Openness Initiative, the staff was increased from 130 people, with a goal of more than 300 people by the beginning of fiscal year 1996.[1] These individuals work page by page, line by line to determine whether a document is properly classified. A decision to declassify a document is customarily reviewed

---

[1] R. Lyons, DOE Office of Declassification, personal communication, February 22, 1995.

DECLASSIFYING DOCUMENTS

and verified by a second individual before its classified marking is removed.[2]

The Committee was told that an experienced declassifier can review approximately 200 pages a day, and we assume that the subsequent reviewer can work twice as fast. If we assume that a declassifier works approximately 240 days a year, reviewing the estimated existing inventory of about 280 million pages of classified documents, at 200 pages per person-day, would require almost 9,000 person-years of effort.[3] The cost of this review, on the assumption of direct and indirect costs per employee of $100,000/year, would be $900 million. This is a large enough sum to warrant careful attention to setting priorities for declassification and serious efforts to develop more cost-effective document review methods. But DOE's estimates of its inventory are uncertain, so these and other cost-and-time estimates must be treated as uncertain as well. Despite the uncertainty of the estimates, there is widespread agreement that quicker, more cost-effective methods must be found.

2. Setting priorities

At present, DOE is only beginning to assess the magnitude of the declassification task it faces. That assessment is an essential first step in formulating plans to address the problem. As part of the Openness Initiative, in June 1994 the Under Secretary of Energy requested heads of DOE headquarters elements, field office managers, and contractors to prepare plans for a systematic review of classified records for declassification as part of a Department-wide systematic declassification review program, including proposed schedules and budgets.[4] It is appropriate to defer commitment to any firm schedule for declassification

---

[2] U.S. Department of Energy Office of Classification, 1991, Order DOE 5650.2B, VI(C)(2b).

[3] Of course, these estimates do not reflect the fact that the inventory of classified documents is growing daily. If the rate of production of classified documents exceeds the rate at which documents are declassified, the inventory at the end of the review of the existing documents would be larger than today.

[4] Curtis, 1994.

review of the relevant collection of classified documents until this information is available. There is still too much uncertainty about the number of classified documents, the cost and effectiveness of alternative declassification methods, the relative urgency of review of different categories of documents, the cost savings achievable by declassification,[5] and the available resources to warrant any commitment to a schedule at this time. The Committee commends the Department's decision to proceed promptly to obtain the needed information. Although the Committee is not in a position to assess the relative priorities for DOE funds, the declassification effort is important and should be accomplished.

DOE should develop better estimates of the direct costs of classification. Creating a classified document imposes a mortgage on the Department to pay for protection of that document and its ultimate review for declassification. At present, DOE has only very aggregated and approximate figures for security costs.[6] Requiring each budgetary unit in the Department to estimate the net current year and long-term costs resulting from classification and declassification actions in that year would provide some incentive to minimize needless creation of classified documents and to expedite declassification.[7]

Since resources for the declassification effort will be limited, priorities will have to be set to determine which documents or classes of documents should be declassified first. Setting such priorities is primarily a policy or value judgment that DOE should make with substantial input from stakeholders. We agree with the Joint Security Commission that **the**

---

[5] The Department believes that its declassification effort has the potential for millions of dollars in savings in the long run, but recognizes that in the near-term costs might increase (Keliher, 1994).

[6] The Department's submission to the Office of Management and Budget (OMB) of its survey of security-related expenditures showed only a single total figure, with no breakdown into the categories requested by OMB (Office of Management and Budget, 1994).

[7] "A formal process should be developed to estimate, as accurately as possible, the direct and indirect costs of classification and security policy" (Meridian Corporation, 1994, p. 92-94).

declassification review process should be driven by customer demand.[8] **A national DOE advisory committee, such as the Information Policy Advisory Committee proposed in this report, could provide advice about declassification of information and documents bearing on national policy debates.**

Information about the health, safety, and the environmental effects of DOE nuclear materials production activities are of particular interest to the communities in the neighborhood of DOE sites. Therefore, establishing priorities for declassification of documents containing such information should have strong local and regional input. Appropriate clearances must be provided to the selected reviewers from the public to enable informed advice.[9]

Another important source of demand for declassification of documents is the historical value of such documents in understanding the nuclear era. The Committee notes that the responsibility for preserving records of historical value lies with the Archivist of the United States, who ultimately must decide which records are to be preserved and which may be destroyed. **DOE should set priorities for declassifying documents of historical value using a process like the one it has already established with the National Archives and Records Administration and stakeholders to deal with documents transferred to the National Archives.**

DOE estimates that there are already on the order of 3 million classified DOE documents in the National Archives.[10] The DOE is

---

[8] "Moreover, given public and congressional concern today that sufficient resources are not being devoted to current FOIA, Privacy Act, and mandatory review requesters, diverting limited available resources to a time-consuming review process that is not driven by customer demand is unacceptable" (Joint Security Commission, 1994, p. 28).

[9] This approach was suggested by a spokesperson for the group of national laboratory directors at the Committee's February 1994 work session in Washington, DC.

[10] DOE estimates that there are approximately 1,000 linear feet of Atomic Energy Commission documents being reviewed at the National Archives by DOE reviewers (DOE Facts, "Declassification of Documents Turned Over to the
(continued...)

supporting the Archivist in declassifying these documents by providing personnel. According to the DOE Office of Declassification, as of February 1995 five full-time declassifiers were working at the National Archives, with a sixth reviewer awaiting assignment.[11] The Department and the National Archives and Records Administration co-hosted a stakeholders meeting in January 1994 to discuss priorities for declassification review with historians, archivists, and researchers. The priority list that the Archives supplied was approved by the stakeholders and is being followed.[12] This process should be continued and expanded. Priorities for declassification of documents of potential historical interest that have not yet been turned over to the National Archives could also be addressed through this mechanism or through the stakeholder advisory bodies at the site or national level, as appropriate.

3. Making the process work

Before launching into a large-scale document review (as opposed to some of the specific steps discussed in this and the next section), DOE should complete the fundamental review of classification policy and revise the guidelines accordingly. Only in this way can DOE avoid having to repeat the review under revised guidelines. In the meantime, DOE should proceed with demand-driven reviews. Once new DOE policy is established (with appropriate stakeholder involvement), a more regular process that gives the public a significant voice in setting priorities should be created.

(...continued)
National Archives and Records Administration," released at Secretary of Energy O'Leary's June 27, 1994, press conference). At approximately 250 pages per inch, this translates to about 3 million pages for 1,000 feet. [This represents at most about 1% of the estimated 300-400 million pages of classified information at the National Archives (Joint Security Commission, 1994, p. 27).]

[11] R. Lyons, DOE Office of Declassification, personal communication, February 22, 1995.

[12] R. Lyons, DOE Office of Declassification, personal communication, February 22, 1995.

As a first step to making the current process work better, DOE needs to expand the base of trained reviewers. Reviewers could be drawn from several sources. Help might be provided by employees and contractors who are currently examining documents for purposes other than declassification -- for example, reviewing documents for dose reconstruction and for waste-characterization studies -- and who might be asked to review them simultaneously for sensitive items. Other sources of knowledgeable reviewers include the reservoir of retired people from all parts of the DOE operations and personnel at the weapons labs whose jobs are being changed or eliminated. Such reviewers would need only a minimum of additional training to be effective in the screening effort.[13]

The plans for systematic declassification review should include planning for production of a record index (with at least a listing of unclassified title, author, date, and document number). Indeed, the early generation of an index could facilitate the overall declassification effort. Stakeholder inputs to priority setting would be greatly facilitated if a simple index of the titles of classified documents were available. Such indexes have been produced at several sites as part of systematic efforts to estimate the radiation doses that have been received by workers and the public over time at those sites. The Committee has been told that those indexes have been very helpful.[14]

Once such indexes are developed, they should be made publicly available so that actual user requests can help set declassification priorities. DOE has taken an important step in this direction by making available through INTERNET a publicly accessible computer database, called OPENNET,[15] that provides bibliographic and locator information on

---

[13] There is a risk that these reviewers could prove to be so steeped in the "old culture" that they resist the new declassification approach. Strong, clear direction and continuing assessment of their work, especially in its early stages, will be necessary.

[14] The value of the index as a research tool would be enhanced if each document in the index were linked to the associated series in a DOE-wide records inventory. This would facilitate historical research by allowing a user to trace a document of interest back to the broader series of which it is a member.

[15] Siebert, 1995.

declassified documents.[16] The Committee commends this initiative and recommends that any record index of documents not yet declassified also be included in OPENNET. DOE should also consider making any such index available on CD-ROM, accompanied by a text search program to facilitate access, and in DOE information centers and major public libraries.

DOE should take steps to ensure that significant documents are not destroyed before they can be reviewed for declassification. Where there is concern that classification may be misused to hide inappropriate or even illegal actions, there will also be concern that important documents revealing such actions could be destroyed before they can be made available for public scrutiny.[17] As noted in Chapter 2, the Federal Records Act (FRA) forbids the arbitrary destruction of federal records, and an agency must secure the approval of the Archivist before disposing of any records. Although the Committee has no information to suggest that documents are being inappropriately destroyed, DOE should ensure that employees and contractors are reminded of their legal obligations under the FRA not to destroy records except as provided in the Act.

Finally, **the Department should assure that declassification and classification decisions are made in a uniform and consistent fashion for both existing and new documents.** Anecdotal evidence suggests that different declassification officials applying the same guidance may reach startlingly different conclusions as to the classified content of a document.[18] The reliability of the entire system is suspect if significant variances are a frequent occurrence. If there is a problem, the solution would seem to rest in improved guidance for declassification officials or perhaps in improved training in applying that guidance. This could be a particularly important step if the entire set of guidelines is revised as a result of the fundamental policy review.

---

[16] R. Lyons, DOE Department of Declassification, personal communication, May 17, 1995.

[17] Oregon Department of Energy, 1994.

[18] Seaborg, 1994.

4. Increasing effectiveness

Looking ahead to hopes for better document management systems, **any computerized system created by the Department should be designed to facilitate declassification of documents and public access to unclassified and declassified documents.** The Committee is not aware of the Department's plans in this regard, but facilitating declassification and public access to information would be a logical component in the design of any such system.[19]

The current process of line-by-line review for large-scale declassification review entails substantial costs and delays caused by limitations in qualified personnel and available funding.[20] **DOE should develop and evaluate faster and more cost-effective declassification methods.** Any expedited process will entail an additional risk of errors -- that is, the unintended release of classified information. Given the concern that it is more damaging to national interests to release classified information inadvertently than to fail to release declassified information, it is not surprising that there are objections to proposals for bulk declassification in which large numbers of documents meeting certain criteria, such as age or inclusion in a particular set of files, would be

---

[19] Segments of the DOE community have significant expertise in the application of computer technologies and, if charged with the task, could no doubt offer numerous suggestions for designing the Department's management system in a fashion that will facilitate the classification, declassification, and handling of the Department's files.

[20] The Hanford Openness Initiative (Oregon Department of Energy, 1994) observed that "[t]he cost to declassify [DOE's classified documents] using today's procedures could be hundreds of millions of dollars and thousands of person-years of labor. The costs both in dollars and labor to do a complete declassification review of Hanford records is unacceptably high. Declassification using present procedures will unacceptably delay public access. New ways to allow public access to this information must be found . . . ."

declassified without individual review.[21] The Committee agrees that such approaches are not promising.

Nonetheless, DOE should investigate screening methods to identify documents that could be subjected to a lower level of review for declassification. Although bulk declassification may generally entail too much risk of inadvertent release of sensitive information, DOE should investigate intermediate approaches in which large quantities of documents are first screened to segregate them into categories according to the likelihood that they contain classified information. This initial screening could be conducted by suitably trained individuals or perhaps by machine using the artificial intelligence approach discussed below. Documents in the categories least likely to contain such information could then be subjected to a less rigorous declassification review, such as by one person instead of two. They could also be given priority for review, since there would be a greater likelihood that such review would lead to declassification of the documents in the lowest risk category. The Committee understands that the Hanford site is currently conducting a large-scale initial screening of this sort (followed by normal declassification reviews) with substantial success.

5. Future improvements

DOE should experiment with artificial intelligence (AI) as a screening tool to identify documents most likely to contain classified Restricted Data. Some explorations of AI applications are already under way.[22] The hope is that use of AI, combined with optical scanning and optical character recognition (OCR) techniques, could permit machine identification of classified information in text or even in stored images and hence greatly reduce the amount of labor involved in the document review. The Committee encourages this effort, but cautions against over-optimistic

---

[21] "...[A]rbitrary bulk or automatic declassification schemes are perceived as risking the loss of information that still requires protection" (Joint Security Commission, 1994, p. 27).

[22] DOE Facts, "Development of Automation to Assist Declassification", released at the June 27, 1994, press conference of Energy Secretary O'Leary.

expectations about savings in time and costs. An AI system might indeed reduce the amount of labor involved in screening documents, but it adds a need for labor to scan documents into the system, edit and correct the initial OCR conversion of scanned text images to standardized computer format, and add necessary document headers. Some records are handwritten and therefore difficult to handle by OCR. It is not clear that the net cost and time required for declassification review alone would be greatly reduced. The most promising application of AI might be as a means of identifying documents with a high probability of containing restricted data (RD), so they can go to the bottom of the priority list for declassification review. This would avoid questions about reliability of AI as a means of ensuring that documents do not contain RD.

If AI proves to be a useful method for screening in declassification reviews, its implementation should be integrated with any broader document management system developed by DOE in order to get double value from the cost of entering documents into computer form for the AI system. Once documents are scanned into an information system and the text images are converted to machine-readable form with OCR software, it is easy to add a bibliographic header and to index the text to allow searches that identify documents of interest to a user. Incorporating the AI declassification review system into a broader document management system would ease user access to declassified materials and would facilitate the public release of documents at the time such documents are declassified.[23]

---

[23] To realize this additional benefit, DOE must have not only the AI technologies to enhance classification review, but also an overall system architecture that can capture a very large number of documents of highly varying quality from a number of sites and make them readily available to users at many other sites. The Committee encourages DOE to evaluate large-scale document management systems already in existence. For example, the Office of Declassification should examine the licensing support system (LSS) being developed by DOE's Office of Civilian Radioactive Waste Management for use in the licensing process for the first geologic repository for high-level radioactive waste. This system is intended to convert 30-40 million pages of documents into machine-readable form, stored as both images and OCR-interpreted ASCII text, and to make this available to a large number of users at different locations over an extended period of time.

As part of its investigation of AI, the Department is seeking to enter into cooperative research and development agreements with the private sector. The Department should not limit this effort to AI, but should open the process up to proposals for innovative methods for quicker and more cost-effective document review, whether or not they involve cutting-edge technology.

In the absence of data about the effectiveness of such approaches, or even about the effectiveness of the current process, it is not possible to evaluate either the cost savings of new approaches to declassification review or the increase in risk of disclosing classified information. Before committing to release large quantities of documents using such methods, DOE should conduct experiments in which various methods are applied to a significant quantity of documents. The error rate should be assessed by the current two-person technique to review the documents that the new method indicates are suitable for declassification. DOE should then carefully reassess the consequences of inadvertent release of classified information of the type that might accidentally be disclosed.

B. Newly Generated Documents

As noted earlier, the Committee has been told that DOE is losing ground on the sheer volume of classified documents: new classified records are being generated faster than others are being declassified. If DOE hopes to achieve a significant reduction in the number of classified documents it must manage and to maintain the inventory at a reduced size, it must take steps to minimize the generation of new classified documents and make those that are created as easy as possible to review for declassification later.

At present DOE does not appear to have a consistent policy to achieve those objectives. The Secretary's directive concerning classification of information related to environment, safety, and health (ES&H) does direct that classification or other dissemination restrictions be used only if they are essential,[24] but such a policy has not been applied generally to all documents. This directive, as well as the UCNI

---

[24] O'Leary, 1993.

guidelines,[25] also encourages segregating classified or otherwise restricted material into an appendix or separable attachment; but again, this does not apply generally to documents containing RD. Exec. Order No. 12,958 requires the portions of documents containing national security information (NSI) to be clearly marked ("portion marking"), but DOE has obtained a waiver of that requirement for DOE documents containing NSI (except for documents that are originally, rather than derivatively, classified, and that contain no RD) and does not use portion marking on RD documents.[26]

To minimize the needless generation of classified documents and to facilitate declassification, DOE should put in place a number of specific procedures:

1) **Classified or otherwise controlled information should be included in documents only if absolutely necessary.** The general rule should be to avoid the future costs inherent in classification wherever possible. The Department could take the language from the Secretary's directive concerning documents dealing with ES&H information and apply it broadly to all documents.

2) **The classifier of new documents should be required to identify the paragraphs of the DOE classification guide requiring the classification action.** The DOE's classification process should require that all newly generated documents be marked in a way that not only facilitates declassification, but also increases accountability and discourages needless or automatic classification. The proposed measure would enhance accountability for the classifying official[27] and facilitate declassification as

---

[25] 10 C.F.R. § 1017.4(b).

[26] The rationale for this waiver is that "it is the position of DOE that the use of classification guides clearly is a superior method for providing guidance to derivative classifiers" [U.S. Department of Energy Office of Classification, 1991, Order DOE 5650.2B, V(C)(5d)]. This does not address the question of ease of declassification.

[27] The DOE's basic classification order already requires documents to be marked

(continued...)

classification guides change, and would be less onerous than requiring a detailed justification for the classification.

3) **Portion marking should be required.** As noted above, paragraph-by-paragraph marking of documents containing NSI to indicate the classified portions is practiced in DOD, but not DOE. Such marking for new documents should greatly decrease the burden of declassification of documents once a decision has been made to declassify an area of information.

4) **Segregating the classified portions should be encouraged.** Classified addenda or tear sheets to separate classified from unclassified information could be used whenever only a small portion of an otherwise unclassified document contains classified information.[28] This would facilitate review and release of the unclassified portions even if the classified portion cannot be released.

5) **Where segregation is not practical, unclassified versions of significant documents of widespread public interest should be prepared.** The Secretary's policy memorandum on ES&H information directs that preparation of an unclassified version be considered if it would allow coherent communication to the public of significant information in these areas. This policy should be extended to other types of documents.

6) **Documents should be coded and indexed so they can be easily tracked, identified, and reviewed for declassification when guides change.** The Department should take advantage of modern computer technology to facilitate the handling of newly created classified documents. Among the opportunities might be the inclusion of coded information with the electronic form of each document that

---

(...continued)
with the name, title, and organization of the authorized classifier [U.S. Department of Energy Office of Classification, 1991, Order DOE 5650.2B, V(C)(1a)].

[28] Joint Security Commission, 1994, p. 18.

preserves a paragraph-by-paragraph justification of the reason for classification by cross reference to the appropriate Classification Guide. Such a system might allow the future rapid declassification of a document (or sections of a document) as the Declassification Guide is modified. Moreover, computer capabilities might be applied to build indices so that a given document or page, which might be found in many files, need be reviewed for classification or declassification only once.

7) **Strict guidance for use of derivative classification should be provided.** Derivative classification occurs when new documents make reference to material in classified documents and must be themselves classified. But, if material in the source documents is no longer classified, then material in the derivative documents should not be classified. The policy should encourage the generators of derivative documents to verify that the content is still classified. The operative guideline should be; When in doubt about classification, check.

8) **Each document, when classified, should carry with it a schedule for declassification review.** Guidelines provided for duration of classification of NSI under Exec. Order No. 12,958 would be useful in this endeavor.

****

Taken together, this combination of broad policy changes, investment in better processes, and specific adjustments in procedures should go a long way to address the currently daunting problems of managing DOE's classified document holdings.

CHAPTER 6

INCENTIVES AND ACCOUNTABILITY

Under the current system, classifiers make decisions about what information should be held secret in a largely closed setting. The culture of national security has traditionally placed a strong emphasis on the costs of openness and little emphasis on the costs of secrecy in a democratic society. The current system provides incentives in favor of overclassification. The possibly irreversible consequences of a mistaken decision to declassify reinforce the tendency to overclassify.

If openness is to be a Department of Energy (DOE) goal, and if it is to have effects at all levels of the organization, the incentive structure must be changed to balance the incentives for overclassification by including rewards for increased openness and minimized use of information control and to have effective systems of accountability to discourage inappropriate controls. If the rewards do not change, the bias will be toward secrecy, no matter what the policy is at the top. Changing the course of an agency with great momentum requires continued application of force in the desired direction, once that direction has been chosen by leadership and communicated to the organization.

A. Steps to Change the Culture

Individuals and organizations often respond better to the promise of rewards than to the threat of penalties, and they tend to produce the things for which they are being rewarded. **DOE should include explicit measures of openness in performance measures for agency personnel and contractors.** Provision of explicit performance measures of openness could be a useful step in establishing concrete positive incentives for openness.

The Office of Declassification should have the enhancement of public access to information as a primary responsibility, rather than as a secondary task that competes with other, more traditional goals. The office should ensure expeditious responses to information requests;

review, evaluate, and investigate complaints regarding access to classified information; provide a focus for public involvement and participation in classification issues; and ensure that recommendations from the public are given a fair hearing in internal deliberations. Complaints will probably relate to timeliness of responses to document requests under the Freedom of Information Act (FOIA), and the office would need to work closely with the offices within DOE responsible for FOIA responses. The office would complement the work of other offices by focusing on classification and declassification issues.

A potentially important means for introducing accountability into the classification system has apparently not been tried on any significant scale. **DOE should require a substantive justification, in terms of explicit criteria, for keeping an area classified whenever it is subject to declassification review.** This change would require classifiers to articulate a substantive justification for withholding a category of material from the public. If the reason for classification is made publicly available, there can be debate within interested communities as to whether the classification decision is in fact justified. (Of course, in some circumstances the acknowledgment that a fact is classified would serve to reveal the secret. No public disclosure of such information, which we understand is rare, would be required.) The justification can also serve as a starting point for judicial review, discussed below, by providing the agency's rationale for classification. The justification could be published for discrete categories of information (for example, the security justification for classifying the number of warheads in the nuclear stockpile), perhaps as part of publicly available classification guides.

DOE should seek advice on important classification decisions from the Information Policy Advisory Board. This Board should render nonbinding but public recommendations to DOE concerning the justification for classifying specific categories of information or for deciding not to declassify information in response to a request for declassification. The Department could choose to reject the Board's advice, preserving its ultimate authority to decide upon classification matters, but political accountability would flow from a decision by the Department to reject the public recommendation of the committee. A board comprising individuals from a range of backgrounds in matters related to nuclear security could presumably apply the expertise that the judiciary lacks, thus overcoming the inhibition that deters aggressive review by the courts.

B. The Issue of Judicial Review

The Commission on Protecting and Reducing Government Secrecy or other appropriate independent group should consider legislation to ensure that meaningful judicial review is available for data classified under the Atomic Energy Act (AEA), as it is for national security information (NSI). It should be recognized, however, that judicial review has significant limitations in the area of classification and is no substitute for the other measures of accountability discussed in this section.

Under the FOIA, citizens can challenge an agency's determination that information has been properly classified as NSI in the courts. Such judicial review is a valuable deterrent to improper decision making. The fact that a classifier is answerable in court for classification decisions no doubt introduces an element of responsibility and balance into the classifier's decision-making process that might otherwise be lacking. And before a court even rules, the act of bringing suit often serves to persuade an agency to release information it had previously claimed to be classified by causing other officials within the agency to review the decision made by the original classifier.

Despite its real and important value, however, judicial review has significant limitations in the context of classification decisions. First, only some classification decisions may be subject to review. For information classified as NSI under an executive order, the law provides for substantive de novo judicial review of the classification decision.[1] For information

---

[1] Exemption 1 of the Freedom of Information Act allows agencies to withhold from the public documents that are "(A) specifically authorized under criteria established by an Executive order to be kept secret in the interest of national defense or foreign policy and (B) are in fact properly classified pursuant to such Executive order" [5 U.S.C. § 552(b)(1)] (emphasis added). The underlined language specifically authorizes a reviewing court to evaluate the substantive adequacy of the classification decision. Goldberg v. United States Department
(continued...)

classified as restricted data (RD) or formerly restricted data (FRD) under the AEA, however, it is not clear that substantive judicial review of classification decisions is available.[2] Instead, the AEA specifically provides for wide agency discretion in determining what information may be kept from the public.[3] Because judicial review is such an important accountability mechanism, we support amending either the AEA or the FOIA to ensure that meaningful judicial review is available for all categories of data classified under the AEA, just as it is available for information classified pursuant to the governing executive order. We see no reason why classification decisions concerning atomic energy related information should enjoy a privileged status in this regard.

Even if the AEA or FOIA were amended in this way, a second, more difficult problem with judicial review remains. Where Congress has specifically provided for review of classification decisions, courts have proven hesitant to exercise this right vigorously.[4] In practice, courts are extremely reluctant to second-guess agency classification decisions and

---

(...continued)
of State, 818 F.2d 71, 76-77 (D.C. Cir. 1987), cert. denied, 485 U.S. 904 (1988).

[2] Exemption 3 of the Freedom of Information Act allows an agency to withhold information that is "specifically exempted from disclosure by statute . . ., provided that such statute (A) requires that the matters be withheld from the public in such a matter as to leave no discretion on the issue, or (B) establishes particular criteria for withholding or refers to particular types of matters to be withheld" [5 U.S.C. § 552(b)(3)]. Unlike Exemption 1, Exemption 3 does not explicitly provide for substantive judicial review of the adequacy of classification decisions. The AEA provisions for restricted data and formerly restricted data, codified at 42 U.S.C. § 2162, likely constitute a withholding statute within the meaning of this exemption. See Virginia Sunshine Alliance v. NRC, 509 F. Supp. 863 (D.D.C. 1981) [(holding a separate section of the AEA, 42 U.S.C. § 2167, falls within Exemption 3), aff'd, 669 F.2d 788 (D.C. Cir. 1981)].

[3] 42 U.S.C. § 2162. "Given the breadth of the Restricted Data concept, it therefore is not surprising that one expert characterized the term Restricted Data as including 'virtually all atomic energy information which the AEC believes warrants protection in the interest of security'" (Cheh, 1980).

[4] Anonymous, 1990.

hence they tend to defer to the agency's judgment concerning what information is properly classified. This judicial deference is rooted in the understandable skepticism of judges that they have the expertise necessary to review decisions made by agency classifiers.[5]

Because of these problems, judicial review alone is not enough to accomplish an accountable DOE classification system. The Department should proceed with the changes discussed earlier, which are within DOE's power now, independent of any need for legislative action.

Finally, DOE should clarify the obligations of members of academic and industrial communities who hold security clearances. American security derives strength from the involvement of individuals outside of government. Such individuals perform important services, providing independent analysis on government programs and technical expertise. Cleared members of the industrial and academic communities frequently make presentations, write papers, or carry out other professional activities on open topics that are contiguous to those currently classified. It is sometimes argued by agencies with authority over the related classified topics that those individuals are accountable to them for ensuring that their work does not include classified information. To make these individuals' obligations free from ambiguities, we suggest that expedient consultation on classification matters should be made available freely to them. If no classified briefings have been provided in direct support of the work in question, no prior review by any governmental agency should be required.

---

[5] One court observed: "[C]ourts accord substantial weight to the determination of Executive Branch officials that information is properly classified . . . [E]ven though the Government has the burden of proving de novo that any information it has withheld fits under one of the exemptions to the FOIA, . . . in the national security context that burden is relatively light. . . . [T]he primary focus of any challenge to a decision to withhold information as classified is normally on the sufficiency of the description of that decision, rather than upon its reasoning." (National Security Archive v. FBI, 759 F.Supp. 872, 875 (D.C. Cir. 1991)(internal citations omitted)). See also Abbotts v. NRC, 766 F.2d 604 (D.C. Cir. 1985) (articulating deferential standard of review in case involving nuclear-related information).

CHAPTER 7

SUMMARY OF RECOMMENDATIONS

The principal recommendations of the Committee, explained in the earlier chapters, are summarized below.

A. Basic Principles

- The goal of the Department of Energy's (DOE) current internal review should be to minimize the subject matter areas that are classified. DOE should seek to construct "high fences around narrow areas" -- that is, to maintain very stringent security around sharply defined and narrowly circumscribed areas, but to reduce or eliminate classification around areas of less sensitivity.
- DOE should shift the burden of proof from the proponents of declassification to the proponents of continued classification.
- In deciding whether a given subject area should be, or should remain, classified, the Department should require that the benefits of classification clearly outweigh the costs.

    - DOE's current criteria for reviewing information for possible declassification should be expanded to include explicitly the benefits of openness in enabling an informed public debate on public issues, and, more generally, in enhancing the public's right to know what its government is doing.
    - Public availability of information should be an important consideration, although this factor should not be the prevailing or overriding criterion for declassification decisions.

- DOE's goal should be "open policies openly arrived at." To the maximum extent possible, the debates about new

information control policies should be open to the public, with ample and credible opportunities for public inputs.
- DOE should establish an Information Policy Advisory Board, appointed by the Secretary and composed of experienced outside experts broadly representative of the major stakeholders in DOE's classification policy. The Board would initially provide systematic external input to the current fundamental review. Later it could serve a variety of functions, for example, making recommendations on priorities for document declassification efforts.
- DOE should take further actions in connection with the fundamental review of classification policy.

    - When the joint review is completed, DOE should indicate publicly which areas of information it believes no longer require protection as restricted data (RD).
    - DOE should promptly release a final version of its report entitled "Public Guidelines to Department of Energy Classification of Information."

B. Priorities for Legislative and Regulatory Changes

- DOE should proceed to use its independent authority to reform its policies and practices toward declassification. DOE should also continue to take the lead in seeking declassification of information about nuclear weapons that it believes can be released without undue risk, but is subject to some degree of control by other agencies.
- DOE should seek legislative authority to simply transclassify to national security information (NSI) any RD that no longer warrants special protection as nuclear-related information but still may be sensitive for other military or diplomatic reasons, thus permitting elimination of the category of formerly restricted data (FRD).
- DOE should not wait for amendments to the Atomic Energy Act (AEA) to implement desired openness policies that are

allowed by the Act. Two important policy changes would be

- Establishing a systematic declassification review of existing documents containing RD, based on priorities reflecting public needs and interests, and on available resources.
- Prohibiting abuses of classification or the control mechanisms established for RD or FRD.

• Where possible, DOE should develop and adopt new rules and procedures as regulations issued under the authority of the AEA and pursuant to the Administrative Procedures Act (APA). Specifically, DOE should promulgate a new regulation concerning classification and declassification of RD.

C. Issues in Classification Policy

• DOE should continue to pursue reciprocal exchanges of information with Russia. Specifically, the Department should explore arrangements in which each party to the exchange retains the right to allow or prevent the public release of the information that it is providing to the other party, so that disagreements about whether information should be publicly released do not obstruct mutually beneficial exchanges.
• The focus on improving control and accounting of Russian fissionable materials should not be allowed to delay the release of declassifiable information about American nuclear weapons that is needed to enable informed debate about policies appropriate to the new international conditions.
• The Committee recommends a thorough reexamination of the need for unclassified controlled nuclear information (UCNI) as a special category. The Congressional Commission on Protecting and Reducing Government Secrecy or other appropriate body should reassess UCNI in

the context of a broader review of controls on unclassified information.
- DOE should evaluate the costs and feasibility of either treating sensitive information related to facility security as a special category of NSI or omitting it altogether.
- If UCNI is to continue to be used as the basis for controlling proliferation-related technical information, the critical interpretations that provide the basis for that use should be included in an updated UCNI regulation.
- If DOE concludes that information now encompassed by UCNI should continue to be protected under this scheme, it should prepare a clear and thorough background information document describing and explaining the rationale for the proposed uses of UCNI and a comparison of alternative approaches to achieving the same objectives.

D. Declassifying Documents

- DOE should develop better estimates of the direct costs of classification.
- Customer demand should play a paramount role in setting declassification priorities, particularly while DOE policy is undergoing fundamental review and change. A national DOE advisory committee, such as the Information Policy Advisory Board proposed in this report, could provide advice about declassification of information and documents bearing on national policy debates.
- DOE should set priorities for declassifying documents of historical value using a process like the one it has already established with the National Archives and Records Administration and stakeholders to deal with documents transferred to the National Archives.
- The Department should ensure that declassification and classification decisions are made in a uniform and consistent fashion for both existing and new documents.

SUMMARY OF RECOMMENDATIONS 93

- Any computerized system created by the Department should be designed to facilitate declassification of documents and public access to unclassified and declassified documents.
- DOE should develop and evaluate faster and more cost-effective declassification methods.
- DOE should experiment with artificial intelligence (AI) as a screening tool to identify documents most likely to contain RD.
- To minimize the needless generation of classified documents and to facilitate declassification, DOE should put in place a number of specific procedures:

    - Classified or otherwise controlled information should be included in documents only if absolutely necessary.
    - The classifier of new documents should be required to identify the paragraphs of the DOE Classification Guide requiring the classification action.
    - Portion marking should be required.
    - Segregating the classified portions should be encouraged.
    - Where segregation is not practical, unclassified versions of significant documents of widespread public interest should be prepared.
    - Documents should be coded and indexed so they can be easily tracked, identified, and reviewed for declassification when guides change.
    - Strict guidance for use of derivative classification should be provided.
    - Each document, when classified, should carry with it a schedule for declassification review.

E. Incentives and Accountability

- DOE should include explicit measures of openness in performance measures for agency personnel and contractors.

- The Office of Declassification should have the enhancement of public access to information as a primary responsibility, rather than as a secondary task that competes with other, more traditional goals.
- DOE should require a substantive justification, in terms of explicit criteria, for keeping an area classified whenever it is subject to declassification review.
- DOE should seek advice on important classification decisions from the Information Policy Advisory Board.
- The Commission on Protecting and Reducing Government Secrecy or other appropriate independent group should consider legislation to ensure that meaningful judicial review is available for data classified under the AEA, as it is for NSI.
- DOE should clarify the obligations of members of academic and industrial communities who hold security clearances.

ABBREVIATIONS USED IN THE REPORT

AEA	Atomic Energy Act
AI	artificial intelligence
APA	Administrative Procedures Act
C.F.R.	Code of Federal Regulations
DOD	U.S. Department of Defense
DOE	U.S. Department of Energy
ECI	Export Control Information
ES&H	environmental, safety, and health
FOIA	Freedom of Information Act
FRA	Federal Records Act
FRD	formerly restricted data
JSC	Joint Security Commission
LSS	Licensing Support System
MINATOM	Ministry of Atomic Energy (Russia)
NNPI	naval nuclear propulsion information
NSI	national security information
NRC	Nuclear Regulatory Commission
OCR	optical character recognition
OHA	Office of Hearings and Appeals
OMB	Office of Management and Budget
OTA	Office of Technology Assessment
RD	restricted data
RDA	Records Disposal Act
UCNI	unclassified controlled nuclear information
U.S.C.	U.S. Code
U.S.T.	U.S. Treaty

REFERENCES CITED

Adler, A.R. 1994. Public Access to Nuclear Energy and Weapons Information, p. 86-91 in D.P. O'Very, C.E. Paine, and D.W. Reicher, eds, Controlling the Atom in the 21st Century, Westview Press, Boulder, CO.

Adler, A.R., ed. 1993. Litigation under the Federal Open Government Laws 32, 18th ed.

Adler, A.R. 1992. Nuclear Secrecy at Critical Mass: A Proposal for Broader Public Access to Information Relating to Nuclear Weapons and Nuclear Power. Draft for review for "Controlling the Atom in the 21st Century", a Conference Sponsored by the Natural Resources Defense Council, December 2-4, 1992, Washington, DC. 33 pp.

Anonymous. 1990. Note -- Keeping Secrets: Congress, the Courts, and National Security Information, 103 Harvard Law Review. 906, 906-909.

Cheh, M.M. 1980. The Progressive Case and the Atomic Energy Act: Waking to the Dangers of Government Information Controls, 48 George Washington Law Review 163, 171.

Curtis, C.B. 1994 (June 27). Implementation of a Comprehensive Fundamental Review of the Department of Energy's Classification Policy. Memorandum from Under Secretary of Energy to Heads of Headquarters Elements and Managers, DOE Operations Offices, Washington, DC.

Green, H.P. 1981 (December). Born Classified in the AEC: A Legal Perspective. Bulletin of the Atomic Scientists, v.37, no. 10, pp. 28-30.

Gusterson, H. 1992. Testing Times: A Nuclear Weapons Laboratory at the End of the Cold War. Ph.D. Dissertation. Stanford Univ., Stanford, CA. 457 pp.

Hewlett, R.G. 1981 (December). Born Classified in the AEC: A Historian's View. Bulletin of the Atomic Scientists, v. 37, no, 10; pp. 20-27.

Information Security Oversight Office. 1994 (March 23). 1993 Report to the President. Washington, DC. 34 pp.

Joint Security Commission. 1994 (February 28). Redefining Security - A Report to the Secretary of Defense and the Director of Central Intelligence. Washington, DC. 158 pp.

Keliher, J.G. 1994 (March 15). Letter to L. Paneta, Office of Management and Budget, in Cost Estimates for Classification Related Activities. Office of Management and Budget, Washington, DC.

Meridian Corporation. 1992 (July 4). Classification Policy Study. Report prepared for the U.S. Department of Energy Office of Classification, Washington, DC. 198 pp.

Morin, N.C. 1994 (February 16). Declassification of Documents Relevant to State Studies. Copies of viewgraphs from presentation at Committee's Work Session.

National Academy of Sciences Committee on International Security and Arms Control. 1994. Management and Disposition of Excess Weapons Plutonium. National Academy Press, Washington, DC. 275 pp.

Office of Management and Budget. 1994 (March 31). Cost Estimates for Classification Related Activities. Washington, DC.

Office of Technology Assessment. 1993 (September). Dismantling the Bomb and Managing Nuclear Materials. Report OTA-O-572, U.S. Government Printing Office, Washington, DC. 202 pp.

O'Leary, H.R. 1993 (June 25). Classification of Departmental Information Relating to Environment, Safety, and Health. Memorandum from the Secretary of Energy to all Department Elements and Directors, Department of Energy Laboratories, Washington, DC. 2 pp.

Oregon Department of Energy. 1994 (March 23). Hanford Openness Initiative. Salem, OR. 7 pp.

President's Blue Ribbon Task Group on Nuclear Weapons Program Management. 1985 (July). Report, Washington, DC.

Seaborg, G.T. 1994 (June 3). Secrecy runs amok. Science, v. 264, no. 5164, pp. 1410-1411.

Shinn, A.M., Jr. 1990. The First Amendment and the Export Laws: Free Speech on Scientific and Technical Matters. The George Washington Law Review, v. 58, no. 2, pp. 368-403.

Siebert, A.B. 1995 (May 7). Success of OPENNET. Memorandum from the Director, Office of Declassification. Washington, DC.

U.S. Congress House of Representatives. 1994. Foreign Relations Authorization Act, Fiscal Years 1994 and 1995. H.R. 2333, 103rd Congress, 2d sess.

U.S. Department of Energy. 1994 (January 25). Balancing Openness and National Security: Proposed Amendments to the Atomic Energy Act -- National Academy of Sciences Committee Version. Draft paper, Washington, DC. 27 pp.

U.S. Department of Energy Office of Classification. 1993 (January). Unclassified Controlled Nuclear Information (UCNI): General Guideline GG-3. Washington, DC.

U.S. Department of Energy Office of Classification. 1991 (December 31). Order DOE 5650.2B: Identification of Classified Information. Washington, DC. [including Change 1 of March 27, 1992, and Change 2 of April 28, 1993]

U.S. Department of Energy Office of Declassification. 1994a (June 27). Draft Public Guidelines to Department of Energy Classification of Information. Draft paper, Washington, DC. 57 pp.

U.S. Department of Energy Office of Declassification. 1994b (January). The Evolution of Atomic Energy-Related Classification and Declassification Policy from World War II to the Post-Cold War Era: An Overview. DOE paper, Washington, DC. 12 pp.

U.S. Department of Energy Office of Nuclear Reactors. 1990 (July 31). Order DOE 5630.8A: Safeguarding of Naval Nuclear Propulsion Information. Washington, DC.

U.S. Department of Energy Office of Security Affairs. 1992 (June 8). Order DOE 5650.3A: Identification of Unclassified Controlled Nuclear Information. Washington, DC.

U.S. Department of Energy Office of the Press Secretary. 1993 (December 7). Energy Secretary Unveils Openness Initiative. DOE News Release R-93-254, Washington, DC. 2 pp.

Zimmerman, P.D. 1993 (February 18). Iraq's Nuclear Achievements: Components, Sources, and Stature. Congressional Research Service, The Library of Congress Report for Congress 93-323 F, Washington, DC. 35 pp + 2 append.

APPENDIX A

CHARGE TO THE COMMITTEE

Appendix A

The Secretary of Energy
Washington, DC 20585

February 2, 1994

Dr. Bruce M. Alberts
President
National Academy of Sciences
2101 Constitution Avenue, NW
Washington, D.C. 20418

Dear Dr. Alberts:

The President is committed to a Federal Government that is more open, and the Department of Energy has taken the initial steps to lift the veil of secrecy surrounding many of the Department's programs and operations. We seek your assistance on this important initiative as we move forward. On December 7, 1993, I announced the declassification of significant amounts of information previously withheld from the public for reasons of national security. A review of departmental classification and information control policies has been initiated to ensure they are in step with the Post-Cold War national and international environment. The National Academy of Sciences can help us in this important process.

We are proposing that the National Academy of Sciences and the Department of Energy jointly sponsor a workshop on the Department's information control policies. The workshop would seek stakeholder views and serve as an educational event at the Department of Energy. One of the primary goals of the workshop would be to examine our classification policy and procedures and prioritize subject areas requiring review for declassification/ public release.

The reputation of the National Academy of Sciences for thoroughness and objectivity is well known. I believe this joint endeavor will benefit the Department of Energy, the scientific community, and the Nation.

I look forward to hearing from you.

Sincerely,

Hazel R. O'Leary

cc:
C. Anderson, National Academy of Sciences
R. Andrews, National Academy of Sciences

APPENDIX A: CHARGE TO THE COMMITTEE

NATIONAL ACADEMY OF SCIENCES

2101 CONSTITUTION AVENUE, NW   WASHINGTON, D.C. 20418

OFFICE OF THE PRESIDENT

February 28, 1994

The Honorable Hazel R. O'Leary
Secretary
United States Department of Energy
1000 Independence Avenue, S.W.
Washington, DC 20585

Dear Secretary O'Leary:

Thank you for your letter dated February 2, 1994. We were pleased to share with the Department of Energy the sponsorship of the February 16 Work Session. I hope you found the feedback at that meeting useful; on our side, the NAS/NRC Committee on Declassification will continue to review the information gathered from the Work Session and the further material you and your staff so generously offered to furnish us, so as to provide you with the committee's findings in a timely manner.

We appreciate the opportunity to help you in this vital component of your, and the administration's, drive for openness.

Sincerely,

Bruce Alberts
President

APPENDIX B

BIOGRAPHICAL SKETCHES OF COMMITTEE MEMBERS

Richard A. Meserve, the committee chair, is a partner with the Washington, D.C. law firm of Covington & Burling, where his practice focuses on environmental and nuclear-related issues. He formerly served as legal counsel to the President's Science Adviser and as clerk to Supreme Court Justice Harry A. Blackmun. He has chaired a variety of National Research Counsel committees, including committees concerned with health, environmental, and safety issues in the DOE weapons complex. Dr. Meserve has a J.D. from Harvard Law School and a Ph.D. in applied physics from Stanford University.

Dean E. Abrahamson, professor at the Humphrey Institute of Public Affairs, University of Minnesota, received an M.A. in physics from the University of Nebraska in 1958, and M.D. and Ph.D. degrees from the University of Minnesota in 1967. He is a trustee of the Natural Resources Defense Council. Dr. Abrahamson's research interest is the intersection of energy and environmental policies, with emphasis on renewable and nuclear energy supply systems. He has been involved with nuclear policy matters, in the U.S. and northern Europe, since the late 1960s.

Lynda L. Brothers, a partner in the Seattle office of the national law firm of Davis Wright Tremaine, specializes in environmental, natural resource, energy and administrative law. She received a B.S. in genetics from the University of California, Berkeley, and an M.S. in biology from the University of Virginia, Charlottesville. She was Deputy Assistant Secretary, U.S. Department of Energy, 1978-1980, and Assistant Director, Washington Department of Ecology prior to entering private practice. Her law practice deals with the regulation, transportation, and disposal of radioactive, hazardous, and solid wastes as well as regulation of water and air emissions.

Thomas A. Cotton, vice president of JK Research Associates, Inc., received a B.S. in electrical engineering from Stanford University; an M.S. in philosophy, politics, and economics from Oxford University; and a Ph.D. in engineering-economic systems from Stanford University. He is a principal in JK Research Associates' activities in the area of radioactive waste management policy and strategic planning. Before joining JK Research Associates, he dealt with energy policy and radioactive waste management issues as an analyst and project director during nearly 11 years with the Congressional Office of Technology Assessment. His expertise is in public policy analysis and strategic planning.

Paul P. Craig, professor of engineering emeritus in the Department of Applied Science, College of Engineering, University of California at Davis and chair of the Environmental Policy Area of Emphasis of the UC Davis Graduate Group in Ecology, received his B.A. in mathematics and physics from Harvard College in 1954 and his Ph.D. in physics from CalTech in 1959. His current interests to environmental policy decision making in area with strong technical components, with special attention to factors affecting institutional and individual credibility.

George A. Ferguson, emeritus professor of engineering at Howard University, Washington, D.C., received B.S. and M.S. degrees in nuclear physics in 1947 and 1948 respectively, and a Ph.D. degree in solid-state physics in 1965 from the Catholic University. His research interests included structure diffraction techniques. He is currently active as a member of the Atomic Safety and Licensing Board Panel of the U.S. Nuclear Regulatory Commission.

APPENDIX B: BIOGRAPHICAL SKETCHES

H. Jack Geiger, the Arthur C. Logan Professor of Community Medicine at the City University of New York Medical School, received an M.D. degree from Case-Western Reserve University School of Medicine in 1958, an M.S. Hyg. degree in epidemiology from the Harvard School of Public Health in 1960, and completed his clinical training in internal medicine in 1964. He subsequently chaired the departments of community medicine at Tufts Medical School and SUNY/Stonybrook. His professional activities have included the initiation and development of the community health center network in the U.S., research on the occupational epidemiology of low-dose radiation exposures, and research and implementation of civil rights and human rights issues in medical care.

Michelle Stenehjem Gerber was born in Schenectady, New York, and received her education at institutions of the State University of New York. She received a B.A. in sociology in 1970 from Cortland State College, an M.A. in history in 1971 and a Ph.D. in history in 1975, both from the State University of New York at Albany. She has worked for several state, municipal, and private historical agencies and has taught history classes at four colleges and universities. She currently is the principal historian at Westinghouse Hanford Company in Richland, Washington, and an adjunct faculty member at Washington State University, Tri-Cities Branch. She is the author of <u>On the Home Front: The Cold War Legacy of the Hanford Nuclear Site</u> (Lincoln: University of Nebraska Press, 1992), an earlier book on American entry into WWII, and numerous articles and documents.

Konrad B. Krauskopf, professor emeritus of geochemistry at Stanford University, received a B.A. in chemistry from the University of Wisconsin, a Ph.D. in chemistry from the University of California, and a Ph.D. in geology from Stanford. Principal research activities have included the origin of hydrothermal ore deposits, the structure of granite batholiths, and the distribution of rare metals in seawater. Currently, a major interest is the problem of high-level radioactive waste.

Wolfgang K. H. Panofsky, director and professor emeritus of the Stanford Linear Accelerator Center (SLAC), earned his A.B. from Princeton in 1938 and Ph.D. from California Institute of Technology in 1942 and has received nine honorary degrees. His special fields of interests are X rays and natural constants, accelerator design, nuclear research, high-energy particle physics, and arms control. He served on the President's Science Advisory Committee, the International Union of Pure and Applied Physics, and as president of the American Physical Society. He has received numerous awards, including the E.O. Lawrence Award, the California Scientist of the Year Award, the National Medal of Science, the Franklin Institute Award, and the Enrico Fermi Award. He is a member of the National Academy of Sciences; as a member of the Committee for International Security and Arms Control he was chairman of the study on Management and Disposition of Weapons Plutonium.

Richard B. Setlow, a senior biophysicist and associate director of life sciences at Brookhaven National Laboratory, received an A.B. in physics from Swarthmore College and a Ph.D. in physics from Yale University. He holds honorary degrees in genetics from York University, Canada, and in medicine from the University of Essen, Germany. His research efforts have dealt with the effects of ultraviolet and ionizing radiations on macromolecules, bacterial and mammalian cells in culture, experimental animals, and humans. His current research is on exogenous DNA damage and its repair and their relations to human carcinogenesis. He is a recipient of the Enrico Fermi Award.

Patricia A. Kelsh Woolf is a lecturer in the Department of Molecular Biology, at Princeton University and a member of the board of directors of Cordis Corporation, General Public Utilities Corporation, National Life Insurance Co. of Vermont, Crompton and Knowles Corporation, and several mutual funds in the American Funds group. Her research focuses on scientific communication and the responsible conduct of research, especially in the biological and medical sciences. She has served on the board of the Council of Biology Editors and the Scientists' Institute for Public Information. She was a member of the National Academy of Sciences Committee on Science, Engineering, and Public Policy (COSEPUP) Panel on Scientific Responsibility and the Conduct of Research.

APPENDIX C

AGENDA FOR JOINT NATIONAL ACADEMY OF SCIENCES/DEPARTMENT OF ENERGY WORK SESSION ON DECLASSIFICATION LECTURE HALL, NATIONAL ACADEMY OF SCIENCES MAIN BUILDING
WEDNESDAY, FEBRUARY 16, 1994

8:30am	Registration	
9:00am	Introductory Remarks	Richard Meserve and Jack Keliher
9:15am	Secretary O'Leary's Openness Initiative	A. Bryan Siebert
10:00am	Panel on Openness and National Security, followed by public discussion	
noon	Lunch	
1:30pm	Panel on Openness and Environmental Issues and Public and Occupational Health and Safety, followed by public discussion	
4:00pm	Steps Toward Reform	Richard Meserve and Dan Reicher
4:30pm	Presentation	Secretary of Energy Hazel O'Leary
5:00pm	Work Session adjourns	

PANELISTS

OPENNESS AND NATIONAL SECURITY
Moderator - **Catherine M. Kelleher**, The Brookings Institution, and Vice Chair, NRC Committee on International Security and Arms Control
DOE Perspective - **A. Bryan Siebert**, DOE Office of Declassification
Weapons Maker Perspective - **Carson Mark**, Los Alamos National Laboratory (ret.)
Legal Perspective - **Allan Adler**, Cohn & Marks
Technical Perspective - **Thomas Cochran**, Natural Resources Defense Council

National/International Perspective - **Steven Cochran**, Lawrence Livermore National Laboratory
Media Perspective - **R. Jeffrey Smith**, Washington Post

OPENNESS AND ENVIRONMENTAL ISSUES AND PUBLIC AND OCCUPATIONAL HEALTH AND SAFETY
Moderator - **Chris Whipple**, Kaiser Engineers, and Chair, NRC Board on Radioactive Waste Management
DOE Perspective - **Peter N. Brush**, DOE Office of Environment, Safety, and Health
Technical Perspective - **Dennis Berry**, Sandia National Laboratories
Military Production Network Perspective - **Fred Allingham**, National Association of Radiation Survivors
State Perspective - **Norma Morin**, Colorado Department of Health
Local Community Perspective - **Amy S. McCabe**, Oak Ridge Local Oversight Committee

LIST OF PARTICIPANTS
Wednesday, February 16, 1994

Attendees
Steve Aftergood
Tom Allen (OGDEN Environmental)
Vicki Allen (Reuters)
Bob Alvarez (DOE)
Scott Armstrong (Info. Trust)
Leonard Brenner (TCI)
Linda Brightwell (DOE)
William Burr (National Security Archives)
Drew Caputo (NRDC)
Paul Carroll (US Congress/OTA)
Tom Clements (Greenpeace)
Phil Coyle (OD/Meridian)
Bob DeGrasse (DOE)
W. Gerald Gibson (DOE)
Stan Goldberg
William Happer (Princeton Univ.)
Roger Heusser (DOE)
David Holt
Ann Hopkins
Daryl Kimball (PSR)
Richard Kleiner (Golden Tech. Co.)
Johanna Kollar (DOE)
Paul Laplante (DOE/OD)
Paul Leventhal (NCI)
Michael McClary (DOE)
Sam McDowell (21st Century Indus.)
George McFadden (DOE)
Priscilla McMillan (Harvard Univ.)
Dennis Nelson
Hazel O'Leary (DOE)
Clifton Peters (ONPI)
James Regens (TUMC)
Dan Reicher (DOE)
Steve Schwartz (Military Prod. Network)
David Schwarzbazh (NRDC)
A. Bryan Siebert (Germantown DOE)
Manny Silverman
Will Vitale (ORAU)
Sheryl Walter (National Security Archives)
Andrew Weston-Dawkes (DOE/OD)
Denise Wilson
James Wright (SNL/CA)

Panelists
Allan Adler (Cohn and Marks)
Fred Allingham (Coalition Citizens)
Dennis Berry (Sandia National Labs)
Peter Brush (DOE)
Stephen Cochran (LLNL)
Thomas Cochran (NRDC)
Jack Keliher (DOE)
Catherine Kelleher (CISAC/Brookings)
Carson Mark (LANL)
Amy McCabe (Local Oversight Cmte.)
Normie Morin (Colo. Dept. of Health)
Jeff Smith (Washington Post)
Chris Whipple (ICF Kaiser)

Committee Members
Philip Abelson*
Dean Abrahamson
Lynda Brothers
Thomas Cotton
Paul Craig
George Ferguson
H. Jack Geiger
Michele Gerber
William Happer*
Konrad Krauskopf
Richard Meserve
Wolfgang Panofsky
Richard Setlow
Gary Stern*
Patricia Woolf

(*withdrew before study completed)

Staff
Carl Anderson
Bob Andrews
Gaylene Dumouchel
Dennis DuPree
Terri Jackson

APPENDIX D

COMMITTEE ON DECLASSIFICATION OF INFORMATION FOR THE DEPARTMENT OF ENERGY'S ENVIRONMENTAL AND RELATED PROGRAMS

MEETING LIST

Meeting	Date	Location
1. Declassification Workshop	February 15-16, 1994	Washington, DC
2. Declassification Committee	March 31-1 April, 1994	Irvine, CA
3. Field Office Subcommittee	May 4, 1994	Livermore, CA
4. Declassification Committee	May 16-17, 1994	Washington, DC
5. FOIA Subcommittee	June 13, 1994	Germantown, MD
6. Declass Teleconference	August 17, 1994	DC-Stanford-Seattle
7. Interagency Subcommittee	September 8, 1994	DOD-Washington, DC
8. Declassification Committee	March 15-16, 1995	Washington, DC